Structural Integrity

Volume 18

The *Structural Integrity* book series is a high level academic and professional series publishing research on all areas of Structural Integrity. It promotes and expedites the dissemination of new research results and tutorial views in the structural integrity field.

The Series publishes research monographs, professional books, handbooks, edited volumes and textbooks with worldwide distribution to engineers, researchers, educators, professionals and libraries.

Topics of interested include but are not limited to:

- Structural integrity
- Structural durability
- Degradation and conservation of materials and structures
- Dynamic and seismic structural analysis
- Fatigue and fracture of materials and structures
- Risk analysis and safety of materials and structural mechanics
- Fracture Mechanics
- Damage mechanics
- Analytical and numerical simulation of materials and structures
- Computational mechanics
- Structural design methodology
- Experimental methods applied to structural integrity
- Multiaxial fatigue and complex loading effects of materials and structures
- Fatigue corrosion analysis
- Scale effects in the fatigue analysis of materials and structures
- Fatigue structural integrity
- Structural integrity in railway and highway systems
- Sustainable structural design
- Structural loads characterization
- Structural health monitoring
- Adhesives connections integrity
- Rock and soil structural integrity.

**** Indexing: The books of this series are submitted to Web of Science, Scopus, Google Scholar and Springerlink ****

This series is managed by team members of the ESIS/TC12 technical committee.

Springer and the Series Editors welcome book ideas from authors. Potential authors who wish to submit a book proposal should contact Dr. Mayra Castro, Senior Editor, Springer (Heidelberg), e-mail: mayra.castro@springer.com

More information about this series at http://www.springer.com/series/15775

Natt Makul

Principles of Cement and Concrete Composites

Springer

Natt Makul
Department of Civil Engineering Technology
Faculty of Industrial Technology
Phranakhon Rajabhat University
Bangkok, Thailand

ISSN 2522-560X ISSN 2522-5618 (electronic)
Structural Integrity
ISBN 978-3-030-69601-6 ISBN 978-3-030-69602-3 (eBook)
https://doi.org/10.1007/978-3-030-69602-3

This Springer imprint is published by the registered company Springer Nature Switzerland AG
The registered company address is: Gewerbestrasse 11, 6330 Cham, Switzerland

To my Mom, Thongkum Makul

October 27, 1957

Preface—Principles of Cement and Concrete Composites

Our current knowledge of binders and cementitious materials for construction is based on assessments of the ways in which these have been applied to date and a long history of using what have become standard concrete mixes. The reactions of these building materials were established based on the hydration reaction or lime carbonation forming hydrates during the hardening process when these concretes were first used in the construction industry. Over time, though, it became evident that the hydraulic cement reactions during the hardening process resulted in more resistance to the effects of extreme environmental conditions and more resistance to potential water damage. Hydraulic cement, therefore, is a material that reacts with water and is capable of resisting deterioration over the long term. However, the stability of this kind of cement in the face of water depends on incorporating toxic and harmful materials.

Based on the growing expertise of professionals involved in the direct practical applications of these concretes, further progress has been realized. Further, significant developments have accrued based on the failure of construction materials in given conditions when the field's understanding of the relevant mineralogical and chemical composition of concrete mixes was still rudimentary. However, expertise in this general area has advanced considerably such that there is now a strong understanding of a diversity of conventional concretes in relation to how best to produce them, the properties of the resulting materials, and the extent to which they remain stable in the long term. Improved measuring techniques and new technologies have further advanced our comprehension of the microstructural development and chemical reactions that take place as cementitious materials go through the hardening process. It is critical both to avoid failures caused by utilizing unsuitable raw materials and to take steps to improve the characteristics of concretes for all applications. In regard to the latter point, applications of admixtures and additives are the result of ongoing processes of research and development.

At the present time, there are many different types of cements available for a range of purposes as well as associated products such as limes and gypsums. Special kinds of cement fulfill specific application requirements; for example, ultra-rapid hardening cement is used for repair purposes and for its remarkable resistance to chemical attacks and high temperatures.

We now have a sophisticated understanding of the composition of natural materials, the development of binder and cement, and the microstructures created during the process of hydration—all of which are found on the formation of amorphous hydrated mineral and crystalline phases. However, the fact remains that the determinants of physical features such as strength development can be measured more precisely than is the case for mineral and microstructure development during hydration.

Today, we base all our measurements on a strong comprehension of crystallographic properties and atomic arrangements. We also know a lot about avoiding failure, enhancing the characteristics of concrete mixes, and minimizing both the production of carbon dioxide cement and our use of energy in the manufacturing process. However, it is still the case that there is much more yet to discover. For example, entirely new binders, cement, and cementitious concretes have been invented in recent years and applications of different types of natural materials have advanced and are continuing to evolve. Multiple kinds of cements, secondary fuels, secondary raw materials with new additives and admixtures included in their composition and subjected to carefully calibrate grinding processes have resulted in many different mineral hydration phases.

It is essential that we continue to develop our expertise in relation to producing, determining the properties of, and applying cement and binders through modern approaches. Likewise, it is essential that we share this expertise broadly in the field. However, the majority of these topics are yet to be explored and discussed in any comprehensive way in the literature. This book, therefore, concentrates on key principles of and issues in the development and use of cement and concrete composites, in which fundamental knowledge of mineralogy and chemistry is combined with consideration of the crystallographic features of amorphous and crystalline phases. Further, the potential of the new concretes is indicated by the various applications for which they can be used successfully.

The book consists of six chapters: Introduction, Principles of Quantum-Scaled Cement, Principles of Low-Carbon Cement, Principles of Fiber-Reinforced Concrete, Principles of Reactive Powder Concrete, and Principles of Tailor-Made Recycled Aggregate Concrete. Reference articles on these topics in the extensive field of cementitious materials are incorporated into the book to demonstrate and discuss cement and concrete of composites.

Bangkok, Thailand
January 2021

Natt Makul

Contents

List of Figures

List of Tables

Chapter 1
Introduction to the Principles of Cement and Concrete Composites

Abstract More advanced techniques have been proposed for construction purposes and improvement cement because of the global sustainability demand. Alongside the integrations of positive policies, emission reductions for all infrastructural projects can be achieved when there shall be swift scale-up in the novel cement use. The book explores techniques such as self-consolidating and advanced nano cements, green cement, and steel fiber reinforcements and how they contribute to construction cost reductions and environmental sustainability. In comparison with the traditional cement, improved multifunctional nano-engineered concretes exhibit advanced functionalities. They for example have up to 146% and 76.5% respective compressive and flexural strengths. They also have improved electrical and thermos-mechanical performances with considerable declines in absorption of waters of about 400%. Technologies of modern engineering aim at generating multifunctional and ultra-high performance concrete substances because of the increased demands for cost-effectiveness, sustainability, and durability. Such building materials are marked by long-term performances and advanced mechanical characteristics. Also, they incorporate characteristics that promote different uses making them sustainable for future applications. Advanced concrete composites are important in multifunctional uses such as in chemical and marine exposed environments because of their high corrosion resistance, affordability, high durability, and lightweight nature. Composite materials (combinations of aggregates) offer an in-built mixture of toughness and stiffness with corrosion resistance and lightweight properties. Such materials are obtained from various compositions with different physical and chemical characteristics. Combinations of such concretes give special capability that gives composite materials an advantage over other improvement methods. Major classifications are explained below:

1. Reinforcement-based composites: The first categorization is founded on reinforcements. Examples are particle-reinforced materials, fiber-reinforced materials, and sheet-reinforced materials. Fiber can be taken from synthetic fibers or organic components such as basalts, carbon, and glasses. Particle-reinforced concretes are categorized into dispersion and large particles. One of the largest particle composites is concrete mixture with gravels and sands. Particle-reinforced concretes have the benefits of production ease and low costs, while particle-reinforced concretes do not perform better than fiber-reinforced

N. Makul, *Principles of Cement and Concrete Composites*, Structural Integrity 18,
https://doi.org/10.1007/978-3-030-69602-3_1

composites. Sheet reinforcements comprise glasses. Glass fiber-reinforced concretes are fiber-reinforced concrete forms consisting of alkali-resistant and high-strength glass fibers distributed into composite matrix. An example of concrete composites founded on reinforcements is RPC. It consists of fine grains of silica fumes, quartz, sand, and cement. Also, it has components of steel fibers and superplasticizers.

2. Matrix phase-based composites: The second composite classification is founded on the matrix phases. Matrix phase-based composites include metal matrix composite, ceramic matrix composite, and polymer matrix composite. Ceramic matrix composites, also known as inverse composites, are custom-made to overcome the challenges of brittleness and monolithic ceramics. They consist of fibers of silicon carbide, silicon nitride (SiN), and aluminum oxide (Al_2O_3). Metal matrix composites (MMCs) consist of metallic reinforcement of aluminum (Al), titanium (Ti), magnesium (Mg), and copper (Cu). Polymer matrix composites (PMCs) have matrices as their components scattered with metal fibers, carbon, and glasses.
3. Nanoscale-based composite: The last classification of composite is based on scales. Bio-composites and nanocomposites are the two types. The nanocomposite involves material mixing and improvements at the nanoscales that result in concretes with exceptional qualities. The demands for bio-composite are for ecological sustainability and biodegradability because they can be obtained from fibers of sugar palms reinforced in matrices of sago starches.

Keywords Cement · Concrete · Composite materials

1.1 Introduction

In general, cements are adhesive materials of all types (Akcay and Tasdemir 2010). It is the binding substances utilized in civil engineering constructions and buildings in a narrow sense. Cements of these types are finely powdered. It is set to form a hard mass when mixed with water. Hardening and setting originate from hydrations that are chemical combinations of the cementitious substances with water that yield gel-like materials with high surface areas or submicroscopic crystals (Zajac et al. 2018). Portland cement is the most important of these.

1.1.1 Definition, Characteristics, and Classification

Portland cements basically consist of compounds of calcium oxides (lime) mixed with alumina (aluminum oxide, Al_2O_3) and silica (silicon dioxide, SiO_2). Calcium oxide (lime) (lime) is derived from lime-rich or calcareous raw materials (Beushausen et al. 2014). Other oxides are obtained from clayey or argillaceous materials. Additional raw materials like bauxite (containing hydrated aluminum, Al $(OH)_3$), silica sand, and iron oxide (Fe_2O_3) can be utilized in small amounts to get the desired constituents.

Chalk and limestone are the commonest lime-rich raw material (Wu et al. 2018). Other calcareous raw materials such as shell and coral deposits are also used. The common argillaceous raw materials are estuarine muds, slates, shales, and clays. A compact cement rock and calcareous clay, Marl contains both the argillaceous and calcareous constituents in percentages that at times estimate compositions of cement (Chen et al. 2018). Blast furnace slag that consists majorly of alumina, lime, and silica is another raw material and is combined with calcareous materials of high content of lime. A white clay, Kaolin that has low amount of iron oxides, is utilized as the argillaceous constituent for white Portland cements. Other possible raw materials are industrial wastes like calcium carbonates and fly ashes from chemical manufactures. Compared with that of the virgin materials, their use is small (Wyrzykowski and Lura 2013). Since the acceptable limit in Portland cement is 4–5%, the magnesia (magnesium oxide, MgO) contents of raw materials should be low. Excessive alkalies, metal sulfides and oxides, phosphates, and flouring compounds are other impurities in raw materials that must be strictly restricted.

Gypsum is another important raw material. During grinding, 5% of gypsum is added to the burnt cement clinkers to control the cement setting time (D'Alessandro 2016). In combined processes, Portland cement can also be produced with sulfuric acids utilizing anhydrite or calcium sulfates in place of calcium carbonate.

Portland cements are manufactured by grinding, milling, and proportion of the following raw materials: Gypsum, $CaSO_2.2H_2O$ obtained together with limestone; iron oxides (Fe_2O_3) from fly ashes, scrap irons, iron ores, and clays; alumina (Al_2O_3) from clay, bauxite, and reclaimed aluminum, and silica (SiO_2) from argillaceous rocks, clays, old bottles, and sand. Without the gypsum, the materials are proportioned to generate mixtures with the desirable chemical compositions (Wetzel and Middendorf 2019). The mixture is the blended and ground through wet or dry processes. At 1450 °C, the materials are then fed via kilns to generate clinkers, graying-black pellets. The iron and alumina act as fluxing agents that reduce the silica melting points to 1450 °C. The clinkers are pulverized and cooled after this phase and gypsums added to control setting time. Then, it is milled extremely powder to generate cement as shown in Fig. 1.1.

1.1.1.1 Chemical Compositions

Four main compounds make up Portland cement. They include tetra-calcium aluminoferrites ($4CaO{\cdot}Al_2O_3Fe_2O_3$), tricalcium aluminates ($3CaO{\cdot}Al_2O_3$), dicalcium silicates ($2CaO{\cdot}SiO_2$), and tricalcium silicates ($3CaO{\cdot}SiO_2$) (Demis et al. 2014). These compounds are abbreviated as C_3S, C_2S, C_3A, and C_4AF, in a designated notation

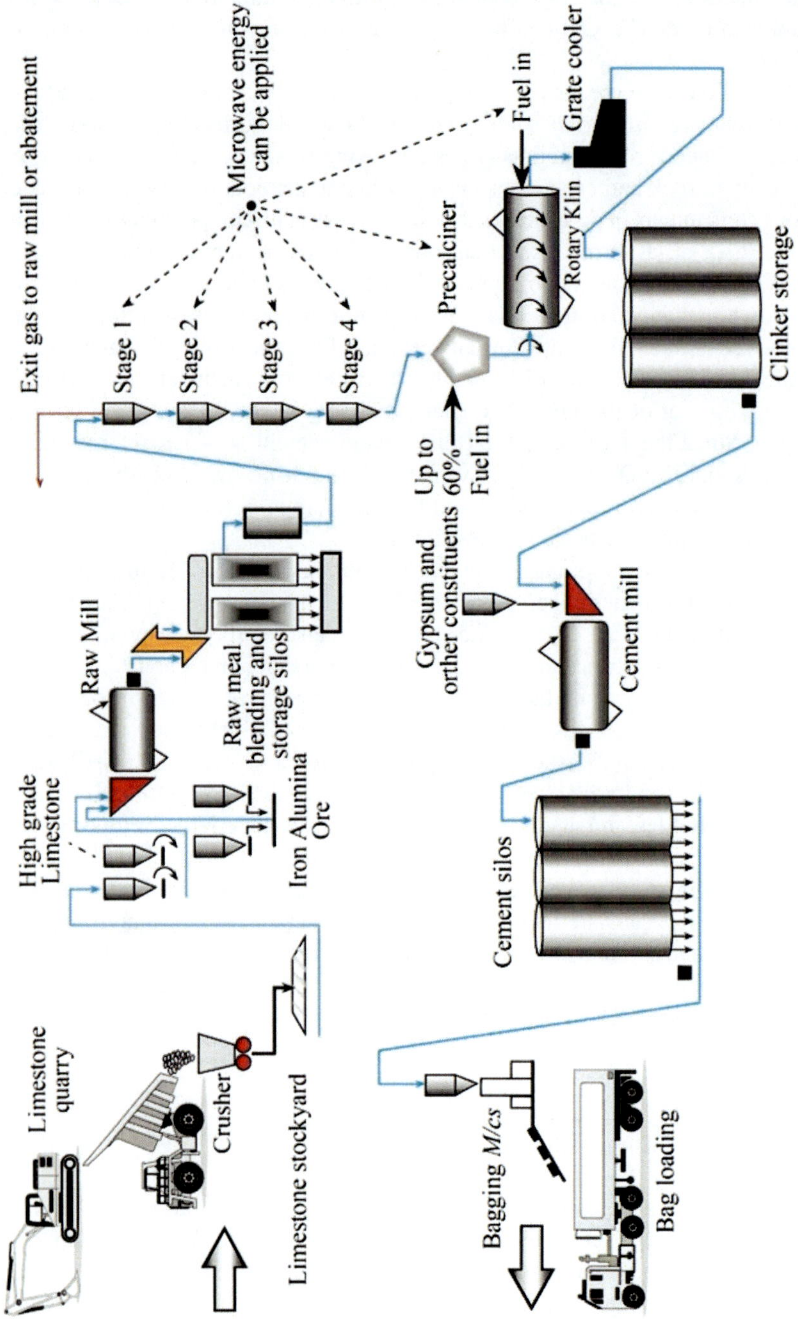

Fig. 1.1 Portland cement manufacturing process (dry process) (Makul et al. 2014)

Table 1.1 Chemical shorthand of cement

Compound	Formula	Shorthand form
Calcium oxide (lime)	CaO	C
Silicon dioxide (silica)	SiO_2	S
Aluminum oxide (alumina)	Al_2O_3	A
Iron oxide	Fe_2O_3	F
Water	H_2O	H
Sulfate	SO_3	$\underline{S}$

differing from the ordinary atomic symbols, in which F stands for iron oxides, A for alumina, S, for silica, and C stands for lime (calcium oxides). Along with minor quantities of other element and alkalies, small quantities of uncombined magnesia and calcium oxides are also present.

1.1.1.2 Chemical Shorthand

A shorthand form is applied in denoting the chemical compounds due to the complicated cement nature as indicated in Table 1.1.

1.1.1.3 Chemical Compositions of Clinker

The cement clinkers produced have following typical compositions. Actual weights vary with cement, and therefore, representative weights only are used. For basic compounds, the shorthand is indicated in Table 1.2.

Table 1.2 Typical compositions of clinker

Compound	Formula	Shorthand form	% by weight
Tricalcium aluminate	$Ca_3Al_2O_6$	C_3A	10
Tetra-calcium aluminoferrite	$Ca_4Al_2Fe_2O_{10}$	C_4AF	8
Belite or dicalcium silicate	Ca_2SiO_5	C_2S	20
Alite or tricalium silicate	Ca_3SiO_4	C_3S	55
Sodium oxide	Na_2O	N	(Up to 2)
Potassium oxide	K_2O	K	
Gypsum	$CaSO_4.2H_2O$	$C\underline{S}H_2$	5

1.1.1.4 Cement Compound Properties

The cement compounds contribute to cement characteristics in various ways.

Tricalcium aluminates C_3A: During the early hydration stages, tricalcium aluminates liberate a lot of heat but have little contribution of strength. C_3A hydration rate is slowed down by the gypsum. Cements low in tricalcium aluminates are sulfate resistant.

Tricalcium silicates, C_3S: These compounds hydrate and harden rapidly. It is majorly responsible for early strength gain and initial set of Portland cement.

Dicalcium silicates, C_2S: These compounds hydrate and harden slowly. Dicalcium silicates are mainly responsible for strengths acquired after seven days.

Ferrites, C_4AF: Ferrites are fluxing agents that reduce the raw materials' melting point temperatures in the kilns to 1435 from 1650 °C (Wang et al. 2020). They hydrate rapidly, but do not contribute much to the cement paste strength. Manufacturers can generate different cement types to suit various construction settings by mixing these materials correctly.

1.1.1.5 Structural Properties

The strength Portland cement develops relies on its finesses to which it is ground and compositions (di Prisco et al. 2013). In the first week of hardening, the C_3S is majorly responsible for the strength developed. For the subsequent increase in strength, the C_2S is responsible. The compounds of iron and alumina that are present only in small quantities make least direct contributions to strengths. Set concretes and cements may suffer degradations from attacks by some artificial or natural chemical agents (Ulm et al. 2010). In seawaters or soil containing sulfate salts, the alumina compounds are the most susceptible to chemical attacks, while the two calcium silicates and iron compounds are more resistant. During the hydration of the calcium silicate, the calcium hydroxide emitted is also susceptible to attacks. Concretes put in large amounts, as in dams can cause temperatures inside the masses to increase as much 40 °C about the external temperatures because cements liberate heat when they hydrate (Guo 2014). Cracking can be caused by subsequent cooling. The greatest hydration heat is exhibited by C_3A, followed in decreasing orders by C_2S, C_3S, and C_4AF.

1.1.1.6 Portland Cement Types

In the United States, the American Society for Testing and Materials (ASTM) standardize five Portland cement types as Type I (ordinary cement), Type II, (modified cement), Type III (high early strength), Type IV (low heat), and Type V (sulfate resistant) (Szajerski et al. 2020). Type III is known as rapid-hardening, and Type II is omitted in other nations. In some European states, Type V is called Type I cement is classified as general purpose cement. Some of its characteristics are that they fairly have high content of C_3S for good early development of strength. Type I is an ordinary Portland cement. They are general purpose cements with no specific characteristics (Ulm et al. 2010). They are applied in general constructions such as precast units, pavements, bridges, and most building. Type II is classified as moderate sulfate resistance cement. They are characterized with low content of C_3S of less than 8%. They are majorly used in structures exposed to water and soils having sulfate ions. Type III cement is categorized as high early strength cement (Han et al. 2019). They are characterized with slightly higher content of C_3S and ground more finely. The applications of Type III cement include cold weather concreting and rapid constructions. Type IV is categorized as slow reacting and low how hydration heat cement. It has low C_3S and C_3A contents of less than 50%. They are used in massive structures like reservoirs. Type V cement is classified at high sulfate resistance cement (Han et al. 2014). Very low content of C_3A of less than 5% is characteristic of Type V cement. Applications of this type of cement include structures exposed to high sulfate-ion levels. Cement can also be classified as ordinary Portland cement and puzzolonic Portland. Ordinary Portland cement is appropriate for foundation constructions and other components. Puzzolonic Portland cements are the best for plastering due to their fineness (Szajerski et al. 2020). Puzzolonic Portland cement can be partially used with ordinary Portland cement.

The following are the various cement types utilized in building and construction projects.

(1) **Rapid-Hardening Cements**
As compared to other types of cements, rapid-hardening cements are very comparable to ordinary Portland cements. Rapid-hardening cements contain finer grinding and higher C_3S contents. As compared to ordinary Portland cements, rapid-hardening cements give higher strength development at early stages. At the age of three days, the strengths of rapid-hardening cements are virtually the same as the strength of ordinary Portland cement at the age of one week with same ratio of water–cement (Wang et al. 2020). The major merit of utilizing rapid-hardening cements is that the formworks can be removed earlier and reutilized in other areas that save framework cost. Rapid-hardening cement can be applied in road works and prefabricated concrete constructions.

(2) **Low Heat Cements**
Low heat cement is produced by reducing the contents of C_3A and C_3S and improving the proportions of C_2S. The initial setting time of this cement is higher than ordinary Portland cement. The cement is also less reactive. It is largely utilized in constructions of mass concretes.

(3) **Sulfate Resisting Cements**
Sulfate resisting cements are manufactured by decreasing contents of C_3A and C_4AF. Cements with such compositions have excellent resistances to sulfate attacks (Song et al. 2018). This cement type is utilized in foundation construction in soils where subsoils contain high sulfate proportions.

(4) **White Cements**
White cements are ordinary Portland cement types that have practically the same strengths and compositions as ordinary Portland cements and are pure white in color. The contents of iron oxides are greatly minimized to obtain the white color (Hakim et al. 2020). The raw materials utilized in these cements are china clay and limestone. White cements are largely used for exterior and interior decorative works such as external rendering of swimming pools, paths of gardens, ornamental concrete products, floorings, facing slabs, and buildings because of its white color.

(5) **Portland Pozzolana Cements**
By either uniformly and intimately blending fine Pozzolans and Portland cement or by grinding pozzolana and Portland cement clinkers with the additions of calcium sulfates or gypsums, the Portland pozzolana cement is made (Harbulakova et al. 2014). It has higher chemical agency attacks than the ordinary Portland cement and generate lower hydration heat. Therefore, concretes made with Portland pozzolana cement are specifically considered best for mass concrete works, constructions in seawaters, and hydraulic works.

(6) **Hydrophobic Cements**
Hydrophobic cements are made by adding mixing ordinary Portland cement and water repellant chemical in the grinding processes. Even during monsoons, the cement stored does not therefore expire (Ulm et al. 2010; Kawashima and Shah 2011; Materazzi et al. 2013). Also, when transported during rain, this cement is claimed to remain unaffected. It is largely utilized for the constructions of water structures such as water retaining reservoirs, spillways, water tanks, and dams.

(7) **Colored Cements**
Colored cements are made by mixing Portland cements and 5–10% mineral pigments during the grinding time. Colored cements is majorly used for exterior and interior decorative works because of the combinations of different colors.

(8) **Waterproof Portland Cements**
Waterproof cements are produced by combining a small proportion of some metal stearates (calcium and aluminum), rapid-hardening cement, and ordinary cements at the grinding time (Harbulakova et al. 2014; Guo 2014). Waterproof Portland cement is utilized for constructions of water-retaining structures such as piers, bridges, dams, swimming pools, retaining walls, reservoirs, and tanks.

(9) **Portland Blast Furnace Cements**
For final grinding, the normal cement clinker in this case is mixed up to 65% of the blast furnace slags (Karakurt et al. 2010; Kawashima and Shah 2011; Kene et al. 2012). Portland blast furnace cements can be applied with advantages in mass concrete works such as constructions in seawaters, retaining walls, abutments and foundations of bridges, and dams.

(10) **Air Entraining Cement**
Air entraining cement is prepared by air entraining agents such as sodium sulfate salts, glues, and resins with ordinary Portland cements.

(11) **High Alumina Cements**
High alumina cements are unique cement produced by reacting limes and bauxites (aluminum ores) at optimum temperatures (Pereira et al. 2012; Qiu et al. 2013; Proske et al. 2018). High alumina cement is commonly referred to as calcium aluminum cements. The compressive strengths of high alumina cements are more workable and every high than ordinary Portland cements. Cements are binder substances that can be classified as non-hydraulic and hydraulic cements, in which cements are used with gravel and sand aggregates to generate concretes in construction sites or with fine aggregates to make mortar for masonry.

(12) **General Application Hydraulic Cements**
General application hydraulic cements are Portland cements that are suitable for general purpose of construction when specific recommendation for other kinds of Portland cement is not required. As a compound of concrete, general use hydraulic cement is used in floors, pavement, bridge, tanks, and pipes manufacturing and construction (Won et al. 2009). This type of cement is very common in Portland cement market due to its generalized usage. The hydraulic cement is used in almost all kinds of construction. It acts as a complement of other types of Portland cements.

(13) **Moderate Sulfate Resistant Hydraulic Cements**
Moderate sulfate resistant hydraulic cements are types of Portland cement used in construction sites where the concentration of sulfate is high, but not severe. The cement neutralizes the reaction of sulfate in concrete buildings. Sulfate elements are found in ground or soil water. In regular structure that is exposed to ground and soil water, moderate Sulfate resistant hydraulic cement is especially recommended (Lamond and Pielert 2006). When sulfate enters into the concrete, it causes a chemical reaction that leads to scaling, cracking and even expansion. The construction may collapse if the reaction is very severe. Seawater concretes are severely affected by the Sulfate. The seawater contains sulfate and chlorides. However, the sulfate reaction with a block of concrete is less severe due to the presence of chloride (Canada).

(14) **Modest Heat of Hydration Hydraulic Cements**
Modest heat of hydration cements are types of Portland cement specially manufactured for the purpose of reducing heat generation. The cement is meant to generate less heat than the general use hydraulic cement. During construction process, it produces heat after reacting with water. The heat generated is referred to as heat of hydration (Aalbers et al. 1994). The cement is recommended in the foundation of broad types of construction. When the mass of the building is large, the moderate heat cement is especially recommended. Thick walls retain heat that may lead to cracking in the future. Breaking and collapsing may occur specifically when the concretes are placed in warm weather.

(15) **Higher Early Strength Hydraulic Cements**
Higher early strength hydraulic cements have similar and physical characteristics of the general use cement. High early strength hydraulic cements are used in constructions that require very quick or fast strengthening. The adhesive can provide the required power during the week. For instance, a high early cement is used in construction of pedestrian paths and pavements (Won et al. 2009). The resins contain fast curing and hence reduce congestion. In situations where the structure put be put into service after a short period, high early strength hydraulic cements are the best. In cold weather, the cement reduces the long period of curing process.

(16) **Lower Lattice of Hydration Hydraulic Cements**
Lower lattice of hydration hydraulic cements is especially utilized in large projects. It is recommended where the rate of heat production, as well as the amount of heat produced by the hydration process, needs to be minimized. In some constructions such as large dams, massive concretes can be used (Won et al. 2009). The temperature rising from such buildings requires to be as minimum as possible.

(17) **High Sulfate Resistant Hydraulic Cements**
High sulfate resistant hydraulic cements are types of cement that gain strength at lower rates than general application hydraulic cements. The cement is used in structures that are exposed to very high sulfate content from ground and soil water (Lamond, and Pielert 2006). Less water that is less permeable to sulfate contents is used in high sulfate resistant hydraulic cement. Little water to construction materials ratio improves the efficiency and effectiveness of high sulfate resistant hydraulic cement.

(18) **White Portland Cement**
White Portland cement obtains its gray color from the raw materials used to manufacture it. The selected materials used in the manufacturing of white Portland cement contain iron and magnesium oxides (Won et al. 2009). The production process is strictly observed to have the final product white in color. The cement is used in building of white structures.

1.2 Fundamentals of Cement and Concrete Composites

1.2.1 Hydraulic Cement

Based on the setting mechanism, cement can be widely categorized into two types:

Cement, on the basis of setting mechanism maybe broadly classified into two types:

1. Non-Hydraulic Cement—sets due to reaction with air
2. Hydraulic Cement—sets due to reaction with moisture

Non-hydraulic cement is that cement which sets as it dries due to the uptake of carbon dioxide in the form of carbonation (Lai et al. 2010; Li and Yang 2017; Klyuev et al. 2018). In other words, this cement reacts with the surrounding air to harden. Hydraulic cement on the other hand hardens when it comes into contact with water (*hydro).* This type of cement is generally used to make concrete which is used in constructing buildings and is also generally waterproof.

Hydraulic cement is the cement that has the capacities of turning compositions such as water, aggregates, sand, and cement into concretes during the wet conditions (Pereira et al. 2012; Pourjavadi et al. 2013; Otero-Chans et al. 2018; Pan et al. 2018). Hydraulic cement is the most common type among all other types. It is majorly applied in constructions cement is added to the mixture for concreting compounds.

Hydraulic cements are products utilized to stop leaks and water in masonry and concrete structures (Lai et al. 2010; Kene et al. 2012; Klyuev et al. 2018). Similar to mortar, it a cement type that sets extremely faster. After it has been mixed with water, it hardens. Hydraulic cement is utilized extensively in the construction sector in scenarios in which structures can be submerged in water or affected by water and in sealing structures below grade (Ohama 2011; Materazzi et al. 2013; Otero-Chans et al. 2018). Hydraulic cement can set in wet surroundings, and therefore, it does need to be dried for it to set. It has many other beneficial properties as compared to other cements.

Some chemical reactions generally occur in the mixtures of hydraulic cement and water. Water containing chemical compounds is formed as a result of these reactions as they help to harden the mixtures (Navak and Kohoutkova 2018). Hydraulic cement is insoluble in water because of these compounds. It is proposed as idle cement for construction of brick buildings.

1.2.1.1 Hydraulic Cement Uses

Hydraulic cement can be utilized above and below grades (Mo et al. 2012). Nevertheless, it is extensively important if utilized in: fountains and cisterns, chimneys, marine applications, sealing around masonry structures and concretes, manholes, basement walls, elevator pits, foundations, drainage systems, and swimming pools.

1.2.1.2 Applications of Hydraulic Cement

Hydraulic cement needs to be used on surfaces that have been washed and are free of grease, oil, dirt, or any other contaminants that can affect the bond with the permanent structures. The steps below are followed for effective applications:

1. Below applying hydraulic cement over the surfaces, all the loose particles should be removed.
2. The AIC recommends that the areas on which the cement will be applied should be undercut.

3. It is recommended to saturate the areas to be worked on for one day before the hydraulic cement is applied.
4. During curing, it is useful to sustain the area temperatures between 32 and 7 °C. Preparation should include avoiding V-shaped cuts and enlarging small holes and cracks.
5. By using mechanical mixers with rotating blades, hydraulic cement should be blended to ensure uniform mixes.
6. Pre-wet mixers and do away with excess water from the mixers.
7. Following the recommendations of the manufacturers, the constructors should add water before adding dry hydraulic cement mix. Once the cement starts to set, no more water should be added.
8. The builders should ensure that they blend at comparatively low speed and that they only blend small amounts of cement that can be placed within working hours.
9. The constructors should not use any other additives and admixtures.
10. The builders should not blend excess water because this can cause segregation and bleeding.

1.2.1.3 Hydraulic Cement Pros and Cons

Hydraulic cement provides some pros even though the cement also has some disadvantages as well. Some of its strengths include.

1. Hydraulic cement offers durable repairs that shall last for longer time periods.
2. The cement hardens and sets faster, often less than five minutes after being combined with water.
3. Hydraulic cement can be painted within sixty minutes of it being applied.
4. Without having to stop the leaking, it can fix leaky basements and pipe.
5. The cement does not shrink.
6. Hydraulic cement is not rusted or corroded.
7. Even if it submerged in water, it maintains its strengths.
8. Hydraulic cement can be used on vertical applications
9. Cold water will retard hydraulic cement, while hot wet accelerates the setting time of the cement.
10. It is cost-effective solutions to the construction industry.
11. Some of the disadvantages of hydraulic cement are as follows:
12. It cannot be used when temperatures are below 4.5 °C.
13. If the temperatures drop drastically within two days or on frozen surface, hydraulic cement will not work.
14. The hydraulic cement only remains workable for ten to fifteen minutes once mixed.

It is important to note that if the problems are because of condensations rather than leaking, hydraulic cement shall not solve the problems, and the builders shall have to apply other solutions.

1.2.1.4 Hydraulic Cement Health and Safety Precautions

The constructors should put on essential personal protective equipment before using hydraulic cement. It should be handled with a lot of caution. The use of masks, gloves, and protective gears is recommended (Akcay and Tasdemir 2010). The users should avoid any contact with skin or eyes and breathing the dust. Inhaling of silica can cause lung complications, even there is no real proof that silica is a carcinogen.

1.2.2 Water–Cement Ratio

In a concrete mix, the water weight ratio to cement weight ratio used is known as the water–cement ratio. It can be described as the water volume ratio to the cement volume utilized in concrete mixes (Kawashima and Shah 2011; Kene et al. 2012). Water plays a great role on the concrete workability and strength especially compressive strength in Fig. 1.2. It directly effects the slump of concrete and hence workability of concrete. It has been found after lots of experiments that there is a specific quantity of water that gives maximum strength for a particular percentage of materials in concrete mixes.

In the concrete strength, a slight change in the amounts of water causes much more variations (Ulm et al. 2010; Ohama 2011). The resultant concretes will be almost dry, may create difficulty in compaction and hard to place in the form if less water is used. Proper setting will not be assured with less water, and therefore, concrete strength gets decreased greatly. On the contrary, water may form honey-combing and larger voids in the set concretes if it is used more as shown in Fig. 1.3. This may decrease the strength, durability, and density of the resultant concrete. Water–cement ratio plays a vital role in generating concretes of required strengths (D'Alessandro

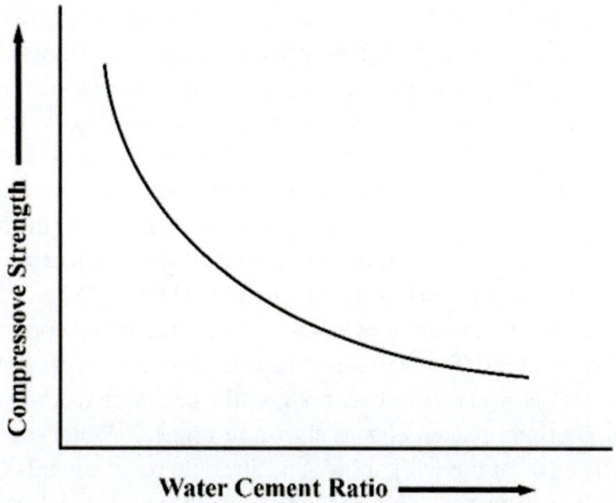

Fig. 1.2 Relationship of increased compressive strength and decreased water–cement ratio

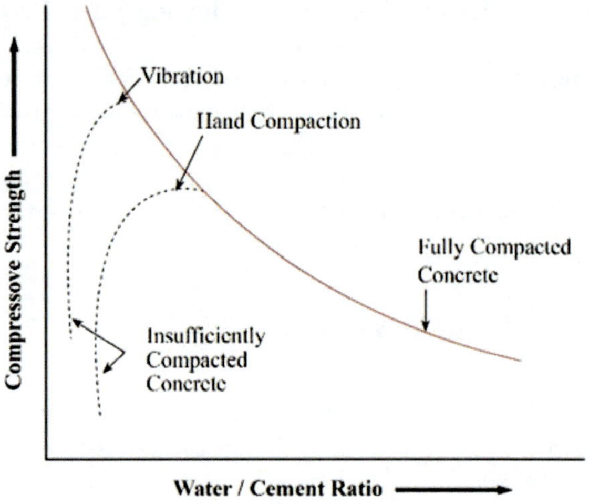

Fig. 1.3 Relationship between strengths and water–cement ratios of concretes

et al. 2016; Zajac et al. 2018). Lower ratios result in greater durability and strength, but can make the mixes hard to work with and form concrete. Superplasticizers and plasticizers can be used to resolve the workability.

Concretes harden due to the chemical reactions between water and cement or because of hydration. The hydration process or reaction between water and cement generates heat referred to as hydration heat (Kene et al. 2012; Klyuev et al. 2018). Around 0.350 L of water are required for each kilogram of mass of cement to completely finish the hydration reaction.

Nevertheless, a mix with a ratio of 0.35 may not flow well enough to be placed or may not mix thoroughly. Therefore, more water may be required to react with cement than the technically needed ratio. Typically, water–cement ratio of 0.46–0.61 is mostly used to develop concretes (Novak and Kohoutkova 2018; Wetzel and Middendorf 2019). Together with plasticizers to improve flowability, lower ratios are used for higher-strength concretes. Concrete strengths depend on the cement paste strengths. The cement past strengths decrease with water and air contents by suitable compaction as shown in Fig. 1.3 and increase cement contents.

In the form S (compressive strength) $= A/B^X$, Duff A. Abrams presented his classic model in 1918; in which B and A are 7 and 96.5 MPa, respectively, and X is water–cement ratios by volume. The water–cement ratio (w/c) which suitably varies from 0.3 to 1.2 by Abrams states that "the strength of concrete is only dependent upon water/cement ratio provided the mix is workable as shown in Fig. 1.4."

In nominal mix concrete such as M10, M15, and M20 concrete constructions, water–cement ratios of 0.45–0.60 are basically applied (Otero-Chans et al. 2018; Pan et al. 2018). Concretes may be combined with water/cement ratios as lower as 0.350. However, this may not give the desirable workability for concrete compactions and proper placements as shown in Fig. 1.5. Water–cement ratios are considered on the basis of the workability and strength requirements for concrete constructions for designed mixes. They put into consideration the free moistures present in coarse

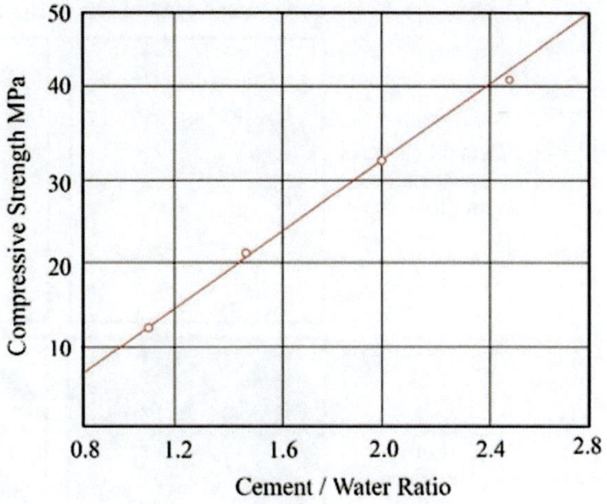

Fig. 1.4 Relationships between compressive strengths and cement–water ratios

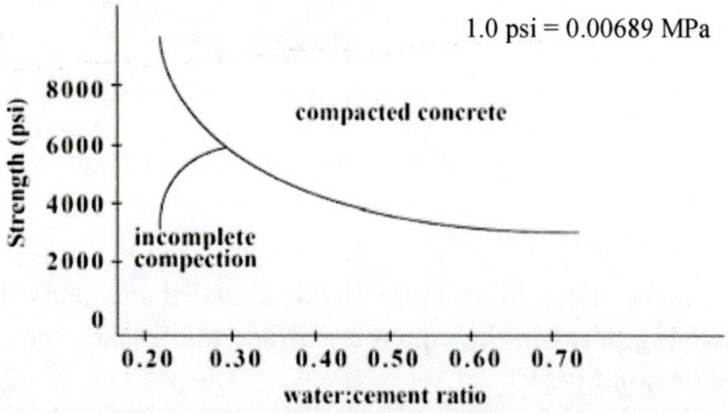

Fig. 1.5 Compressive strength of compacted concrete compared with incomplete compaction

aggregates and sand. For concrete constructions, water ratios are not computed, but they are selected from the different tests of workability appropriate for the concrete type or concrete structural member type (Pourjavadi et al. 2013). Different concrete construction types and different structural members need different workability for correct compaction, placement, transportation, and mixing of concretes.

It is true that the concrete strengths are directly proportionate to the water–cement ratios. Decrease in water–cement ratios increases the strengths of concretes to some extent, but an optimum ratio is to be maintained so that entire cement part of the concrete can take part in reaction, failing of which the concrete will decrease its strength. Moreover for lesser water–cement ratios, the concrete workability will be affected (Song et al. 2018). So that it is difficult to obtain finishing of your shape. On the contrary, if the constructors increase the water–cement ratios, it will reduce the strengths of concretes. To complete the chemical reaction of cement and water, optimum quantity of water is required. More than that, residual quantity of water will

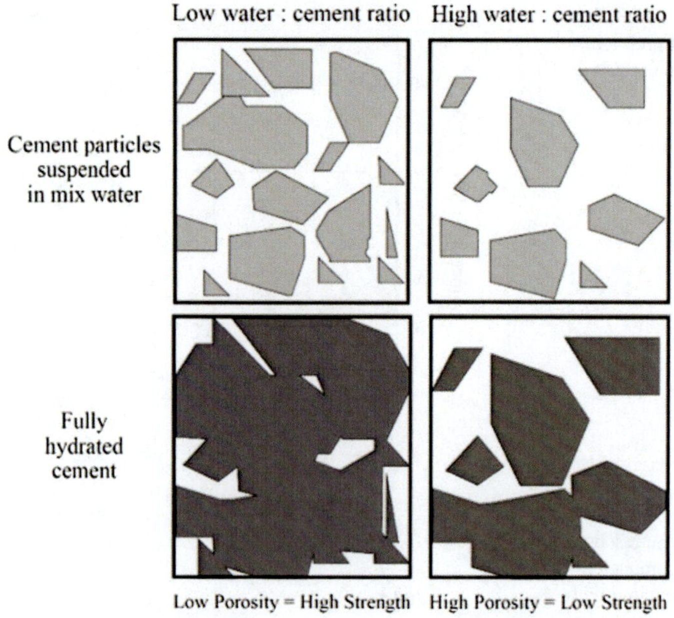

Fig. 1.6 Particle distributions between cement particles suspended in mix water and fully hydrated cement

be in the concrete structure. In due course that water will evaporate from the concrete leaving pores in the concrete surface that makes the concrete of lesser strength as shown in Fig. 1.6.

For the preparation of cement mortar, you require sufficient quantity of water. If the builders add more water, cement will not have any binding property, and if you add less water, there will not be any cohesion in the mortar (Ulm et al. 2010; Wang et al. 2020). So, we have to add correct quantity of water with the cement so that we will have a cement mortar which binds well and get the required strength day by day in the increased manner. Here, the ratios between weights of the cements to that of weights of water are called water–cement ratio.

1.2.3 Hydration Reaction

Portland cements get their strengths from chemical reaction between water and cements (Wetzel and Middendorf 2019). The process is referred to as hydrations. Hydration is a complicated process that is best comprehended by comprehending the cement chemical compositions. Portland cements basically consist of compounds of calcium oxides (lime) mixed with alumina (aluminum oxides, Al_2O_3) and silica (silicon dioxides, SiO_2). Calcium oxides (lime) are derived from lime-rich or calcareous raw materials. Other oxides are obtained from clayey or argillaceous materials. Additional raw materials like bauxite (containing hydrated aluminum, $Al(OH)_3$), silica sand, and iron oxides (Fe_2O_3) can be utilized in small amounts to get the desired constituents.

C_3S, calcium silicates, and C_2S are the most significant hydraulic compositions. The calcium silicate reacts with molecules to form calcium hydroxide ($Ca(OH)_2$) and hydrated calcium silicates ($3CaO{\cdot}2SiO_2{\cdot}3H_2O$) upon mixing with water. The short abbreviations given to these compounds are calcium silicate hydrate (C–S–H) (Kene et al. 2012). The average formula that represents this is $C_3S_2H_3$. The following reactions crudely represent the hydration reactions and CH:

$$2C_3S + 6HgC_3S_2H_3 + 3CH_2C_2S + 4HgC_3S_2H_3 + CH \quad (1.1)$$

As C–S–H and CH are generated, the cement begins to harden following the dormant time that can last several hours (Kene et al. 2012; Klyuev et al. 2018). These are the cementitious materials that bind cements and concretes together. Cement and water are continually consumed as hydration proceeds. The products of CH and C–S–H fortunately occupy approximately the same volumes are the original water and cement. Shrinkage is manageable, and volume is approximately conserved. The formulas above do not all form an ordered uniform composition structure even though the above formulas treat C–S–H as specific stoichiometry with formula $C_3S_2H_3$. Actually, C–S–H are amorphous gels with highly variable stoichiometry (Ulm et al. 2010). For example, the ratio of C to S can vary from 1:1 to 2:1 depending on curing conditions and mix designs.

1.2.4 Pores

Most of the water from hydration is held in pores either as physically or chemically bound water. The physically bound water in the pores evaporates to maintain equilibrium with atmospheric relative humidity (Wyrzykowski and Lura 2013; Wu et al. 2018). As the RH value reduces, water starts evaporating from smaller capillary pores—this is where shrinkage might occur. The chemically bound water (water bound in C–S–H layers) still tends to persist after the concrete comes into an equilibrium with atmosphere. This water and its movement are considered responsible for time dependent effects and most commonly attributed to creep effects in concrete.

To be precise, some amount of water is always present in concrete either as chemically bound water or as in the form of pore solution of whatever it might be—some amount still stays. When the constructors mix water into cement, most of the water is used in the chemical reactions. But after a month or so, the concrete has hardened to the standardized strength, and it will contain pores (Pourjavadi et al. 2013; Pan et al. 2018). These pores can absorb water. Concrete will change continuously during its lifetime, and the maximum strength will occur after about 50 years or so, depending on concrete and climate. So, basic concrete is actually not waterproof. It has cavities, like a very tight sponge. The surplus water, little as it may be, will just vaporize.

Different concrete designs have different levels of permeability. Ultra-high performance concrete is highly impermeable (Wetzel and Middendorf 2019). They have removed the large or coarse aggregate from the mix. The result is a highly impermeable and very strong (158.6 MPa at 28 days) concrete.

Concrete does have air holes in it. In fact, it is often required that there is a minimum air content in the mix. The secret is to have micro-air bubbles rather than large air bubbles. The larger air bubbles tend to provide a weak spot, whereas the micro-air bubbles do not appear to have this affect.

Water is an ingredient in concrete. Water is part of the chemical reaction that takes place to make concrete strong. The process is referred to as hydration. Hydrations are the chemical reactions where the main constituents in cements for chemical bond with molecules of water and become hydration products or hydrates (Wetzel and Middendorf 2019; Wang et al. 2020). The curing process, which includes hydration, never really stops. It just slows down. The hardened concrete may absorb the water more than permeate it.

Porosity is highly relative (Wang et al. 2020). Certainly, it has enough little pores that typical concrete might absorb 5–10% of its weight in water from oven dry. However, in ordinary concrete, the pores are not continuous. Therefore, there is no easy path for water to permeate through. Properly mixed, placed, and finished concrete is not highly porous at the macro level. If it was highly porous, it would reduce the strength. Concrete is permeable, but not excessively permeable. All concrete should be sealed with a membrane forming sealant (Harbulakova et al. 2014). The sealant promotes proper curing and provides protection from chemicals and stains.

1.2.5 *Interfacial Transition Zone*

Water–cement ratio gradients develop around the particles of aggregates in fresh concretes during casting, leading to different microstructures of pastes of surrounding hydrated cements (Kene et al. 2012). These zones around the aggregates are known as the interfacial transition zones (ITZs). During hydration, a greater water–cement ratio means diffusion processes, and these zones can be subsequently defined as heterogeneous areas with complementary gradients and porosity gradients of hydrated and anhydrous stages (Pan et al. 2018). The initial water–cement gradients around the concretes are reduced, and the ITZs are densified by using well-dispersed and fine mineral additions (Wetzel and Middendorf 2019). The ITZ microstructures can be enhanced in calcareous aggregate vicinity that react with calcium aluminate of Portland cement pastes to form calcium carbo-aluminate.

Concretes are considered to be two phase materials: paste and aggregate phases (D'Alessandro et al. 2016). The concrete complexity begins to show up at microscopic levels, especially in the vicinities of large particles of the aggregates. These are areas are considered as third phase, known as the transit zones that represent the interfacial region zones between the hardened pastes and coarse aggregate particles (Kene et al. 2012). The transition zones are planes of weaknesses and thus have great influences on mechanical concrete behavior.

Transition zones are composed of some huge concrete pastes. The paste quality sometimes is poor (Pourjavadi et al. 2013). Water is accumulated below large, elongated, and flaky particles of aggregates because of internal bleeding. This decreases the bonds between the aggregates and pastes.

1.2.6 Strength

Specifications of national cement lay down different test to which cements must conform to control the strength, setting time, soundness, and the finesses of the cement (Wu et al. 2018). The first experiments the authors carried out are about the hydraulic cement fineness. The type of cement the authors used is the Type I of Portland. This Portland cement form is appropriate for general construction, particularly buildings, and bridge (Beushausen et al. 2014). The authors observed from the experiments that we can see the presence of coarse cement particles through sieving methods and measure the material fineness.

Fineness is a cement property that shows its specific surface and particle size (Karakurt et al. 2010). This characteristic indirectly affects the hydration heat. This implies that the surface areas of cement particles should be higher for a higher heat of hydration. Strengths develop on the concretes as hydrations take place. This would mean that the fineness of cement could affect its reactivity with water (Otero-Chans et al. 2018). After following the procedure correctly, we have come up with an acceptable data. In our first trial, the residue that is left in the sieve is subtracted to the original weight which is 100 g, and then, we divided it by the original weight. Fineness of the cement is 97.28%. For the second trial, we did the same procedure, and we got a 97.96% fineness of the cement. With this, we are able to compute the average fineness of the hydraulic cement which was 97.62% (Wetzel and Middendorf 2019). We considered our experiment a success because our professor told us that fineness should not be less than 80%.

1.3 Integrity Problems of Cement and Concrete Composites

Concrete by definition is a composite material consisting of fine (<4.75 mm) and coarse aggregates (>4.75 mm) held together by a binding agent, i.e., cement (Chen et al. 2018). If the builders use just cement and sand, then it would be called mortar. It is advisable to use coarse aggregates.

The strength of concrete is greatly influenced by its porosity. Most concretes and aggregates have high strengths and low porosity. Therefore, the cement paste matric strength controls the strength of the concretes (Kawashima and Shah 2011). One of the significant factors influencing the concrete strengths is water–cement ratio. High water–cement ratios reduce the strength of matrix and increase its porosity. Therefore, it is the aggregates that make concretes strong.

The most significant problem engineers might face with attempting to building anything, however, small it might be without aggregates (coarse/fine) is severe shrinkage cracking (Pan et al. 2018). Cement paste hydrates and releases a lot of heat. Now, imagine if the material is built completely with cement, the heat it generates and the shrinkage it undergoes when water is evaporated. Aggregates alleviate this shrinkage and heat of hydration effects and utilize them as a binder to fill the volume in demand.

1.3.1 Design and Construction Errors

The designs and constructions of structures involve various errors. While other deficiencies play no role, some of the mistakes are crucial in its ultimate failure. Some of the defects also need to be considered as errors only in hindsight because the structure engineering practice and theory are quickly growing at the time (Pourjavadi et al. 2013). Various vital human aspects can contribute to the failures of the structures throughout their designs, constructions, and maintenance.

1. The engineers also may fail to incorporate wells of uplift relief on the parts of sloping supports of the structures. This limitation could fail the buildings and structures.
2. The designers might fail to evaluate the significant structure arch strains that are overstrained in the structures' upper feet. The engineers may be unable to carry out the arch stresses trial load analyses in mass concrete of the structures.
3. The staff of the structure can fail to put into consideration the mass concrete hydration heat. High alkali cement may be used in the mass concrete structures that accelerate their failures.
4. The engineers may fail to remove the laitance layers (high cement pastes of water contents) between the lifts of the concretes. This weakens the walls of the structures.
5. Before incorporations in the concrete of the structures, the team may also fail to wash concrete aggregates.
6. Failures to pound the upstream structure faces to minimize the tensile forces through girder actions are a big mistake. These mistakes are direct consequences of making assumptions that no substantial uplift stresses would develop under the maximum sections of the structure.

7. The failures to offer the structures with plastered contraction intersections demonstrate a high level of negligence on the part of the team that designs and constructs the structures.

The structure of a building is much more than its foundations, floors, and walls (Demis et al. 2014). To sustain long-term stabilities, all constructions virtually depend on various support materials that include layers of cement and soils. Construction materials testing (CMT) is a necessary process that assists constructors and site managers in determining the possible risks and challenges before the commitments of resources to the projects (Klyuev et al. 2018). Also, construction materials testing (CMT) is vital for keeping the structures in line with applicable law demands that include environmental regulations and occupational safety. Construction materials testing (CMT) provides detailed solutions for material testing, including special inspections, field investigations, and laboratory tests. Engineering and materials testing ensures that essential materials deployed during construction meet all the fundamental benchmarks of quality (Pereira et al. 2012). Through an examination of roof materials, structural steel, reinforcing, asphalt, masonry, concrete, soils, and earthworks, construction materials testing (CMT) control the risks related to the construction of a structure.

1.3.2 Low Strength

There are lots of factors which may affect concrete strength (Pereira et al. 2012). Curing of concretes is the processes of sustaining adequate moistures in concretes within proper temperatures range to help hydration of cement at early stages. Various chemicals duration hydrations contribute to setting and hardening. Techniques of proper curing will prevent in-situ concretes from cracking, drying, and shrinking and eventually affect the concrete strength and the structure durability. Strength of concrete depends on materials of what they made. Concrete is made of cement, coarse aggregate, fine aggregate, and water (Yang et al. 2017). For common observation, it is easy to break a rock by tensile force and very hard to break by compression. So, cement is finest particles of rock, fine aggregate again finer particles of rock. Water is used to combine finer rock particles into larger rock. In concrete, small cracks are there, and when tensile forces act, these cracks are widened. Tensile forces can be transferred only through the un-cracked concretes. Concrete is an aggregate comprised of sand, gravel, and cement (usually). As non-metals, they are covalently bonded, and covalent bonds are very strong bonds, especially in giant lattices, meaning it is very hard and strong (di Prisco et al. 2013). Covalent bonds are when atoms share electrons (one from each atom) to help complete their outer electron shell. Sand and gravel have melting points of several thousand degrees, showing that they have very strong bonds as it takes a lot of energy to break them.

1.3.3 Cracking

When cement and water are mixed, they react (Lai et al. 2010). The process is called hydration of cement. This reaction is exothermic in nature. For a significant amount of time (28 days after setting), this reaction is very rapid and accounts for the gain of strength of concrete. After 28 days however, the reaction slows down gradually up to a negligible point (Proske et al. 2018). But as the process takes years to complete, it keeps emitting heat, and sometimes, this leads to the formation of cracks in concrete.

Concrete is basically a homogeneous mixture of cement, sand, and aggregate along with water in fixed proportions (Zajac et al. 2018). The basic ingredient of concrete, cement, acts as a binding material for the other two and thus forms a hard mass. Due to volume change in concrete, cracks are developed. Few reasons of volume changes include internal volume change due to heat of hydration, sulfate action, moisture movement, pozzolanic action, rusting of steel reinforcement and shrinkage (Guo 2014). Probably the most common cause is shrinkage cracking caused by the change in volume as concrete cures, typically handled by saw-cutting contraction joints at appropriate intervals. Delay in cutting those joints will result in cracking. Cracks are found after slab concreting became shrinking. Wet concrete becomes hard due to a chemical process which liberates heat (Li and Yang 2017). This is the reason why the concrete must be cured by keeping the surface that is exposed to air is kept wet.

Failures to provide enough curing to the concrete cause it to crack (Qiu et al. 2013). When concrete is in the process of gaining strength (hardening) after setting, it heats up due to tricalcium silicate's reaction with the water. Concrete uses mixture of cement, fine aggregate, coarse aggregate, and water. Water gets evaporated due to setting of cement generating from heat of hydration. The heat, then, evaporates the water content rapidly (Yang et al. 2017). Since the initial process of concrete gaining strength goes on for years (but most of the strength is developed within 28 days), it requires a lot of water in the initial phase of 14 days. So, if the concrete lacks water when it is mixed, it will start to crack if it is not cured. As the concrete cures, a process that can take several years, it contracts as water evaporates. The contraction causes a tensile stress that pulls the concrete apart, creating the crack (Han et al. 2019). Steel or other reinforcement such as thin fibers is added to concrete to provide tensile strength and control cracking.

Also, concrete, like any other substance, expands on heating and contracts on cooling. During the day, it is subjected to continuous heating, and in the night, the temperature falls well below to provide a cooling effect, depending on the regional weather (Lura and Terrasi 2014). This alternate cooling and heating is also partly a cause of crack formations. Some minute cracks in concrete developed during setting of concrete are sometimes filled with moisture. Due to the freeze and thaw action, as this water expands in the crack, it exerts pressure on the surrounding surface, therefore increasing the crack size. Cracks may also be developed due to applied loads on the structure (Song et al. 2018). This may happen in case when the applied load is near about or well above the permissible values as shown typically in Fig. 1.7.

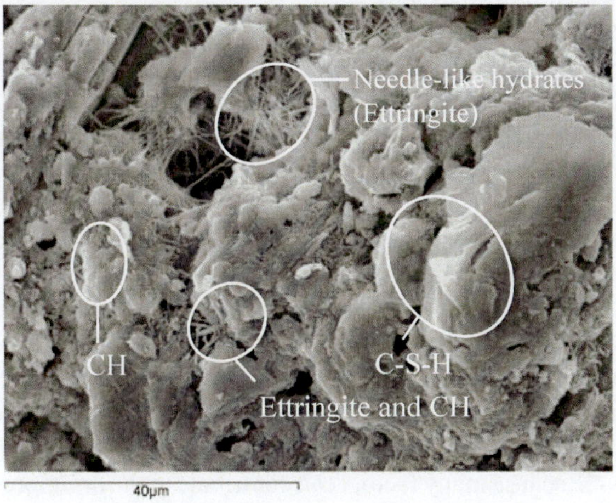

Fig. 1.7 Typical cracking in mortar

1.3.4 Shrinkage

Drying shrinkage is described as the hardened concrete mixture contracting because of the capillary water loss (Ulm et al. 2010). These shrinkages cause increases in tensile stresses that can result in external deflections, internal warping, and cracking, before the concretes are subjected to any form of loading. All concretes of Portland cement undergo hydral volume change or drying shrinkage as the concretes age. In concretes, the drying shrinkage is very vital to the engineers in the designs of structures (Han et al. 2014). Drying shrinkages can take place in foundations, tanks, pre-stressed members, bearing walls, columns, beams, and slabs. Drying shrinkages are dependent upon various factors. Such factors include size of member, dry environments, amounts of moistures while curing, mixing manner, proportions of the components, and the properties of the components. Under normal conditions, concrete cured will undergo some volumetric changes (Materazzi et al. 2013; Novak and Kohoutkova 2018; Otero-Chans et al. 2018). Drying shrinkages take place majorly due to the capillary water reductions by the water and evaporation in the cement pastes. The greater the water amount in the fresh concretes, the higher the effects of drying shrinkages. The shrinkage potentials of specific concretes are influenced by mixing amounts, curing, placements, slumping, temperature fluctuations, and the elapsed time after water addition. Also, the components of concretes are very important. Each cement and aggregate type has clear features, each contributing to concrete shrinkages. The quantities of admixtures and water utilized during mixing also have indirect and direct effects on concrete drying shrinkages. Concrete shrinkages take place majorly because of the mixing capillary water evaporation (Ulm et al. 2010; Wyrzykowski and Lura 2013; Snoeck et al. 2018). The shrinkage severity depends on the concrete physical properties including the surrounding temperatures, the structure location, and structure size.

1.3.5 Discoloration

Surface discolorations are the hue on the surfaces of a single concrete placement or the non-uniformity of color (Hakim et al. 2020). In concretes, calcium chloride is a chief cause of concrete discolorations. Concretes that are not uniformly or properly cured can develop discolorations. Discolorations because of changes in sources of fine aggregates or cement in consequent batches in placement sequences might take place, but is generally insignificant and rare. Cements that have hydrated to greater levels shall basically be lighter in color especially improper timing of finishing, insufficient mixing time, and inconsistent application of admixtures (Kawashima and Shah 2011; Murtazaev et al. 2018; Snoeck et al. 2018). On concretes containing ground slags as cementitious materials, a yellowish to greenish hue may appear. With time, this shall disappear.

Because many factors can contribute to the discolored concretes, discoloration is a tricky subject (Harbulakova et al. 2014). Whether taking place in sequential concrete placements or single placement, discoloration boils down to inconsistency. This can be an inconsistency in workmanships or materials. It is important to sustain consistency throughout the project, particularly in mix proportioning to avoid issues of objectionable color (Novak and Kohoutkova 2018). Mix deliveries that are extremely different from batch to batch make it hard to accomplish a uniform color of concrete surface.

1.3.6 Repair and Maintenance Errors

Construction materials testing (CMT) includes elements of potential impacts on neighboring structures, local terrains, and composition and quality of the soil (Otero-Chans et al. 2018). Also, it involves the examination of woodwork, masonry, steel, and concrete components throughout the structures. The process of testing is vital since it allows builders and supervisors to determine faults and deficiencies before the real tests of stresses, in which environmental and personal safety are at risk. Construction materials testing (CMT) is seriously helpful and cost-effective throughout the process of construction, even though comprehensive inspections are not foolproof (Jones and Grasley 2011). Construction materials testing (CMT) assists in avoiding costly renovations or repair works to mitigate risk and correct faults to other properties and people. Different proof and inspections of material quality are also needed bylaws depending on the structure type and location.

Construction materials testing (CMT) starts before the construction teams even begin excavating to create rooms for foundations in many construction projects (Ohama 2011). The phases of construction materials testing (CMT) include analyses of possible impacts on the nearby structures and terrains as well as detailed examinations of in-place materials such as asphalt, soil, and aggregate. The grounds are tested for compositions and chemistries, so they can be correctly treated or stabilized to reinforce a structure better (Wang et al. 2020). Another essential construction material testing (CMT) component is foundation installation observation and monitoring because using faulty materials can lead to a disastrous and chronic consequence. Also, the engineers and builders carry out periodic examinations at specific points throughout the process of the construction in ensuring that all materials meet the minimum standard (Ulm et al. 2010; Ohama 2011). This involves analyses of construction envelope quality assurances, reinforcing steels, and fireproofing.

1.4 Scope of the Book

The book is developed to reflect present advancements and developments being undertaken in the general fields of technology of cement–concrete composites and in general, in the manufacturing, performance, and applications of cement-oriented building materials. The book consists of six chapters: introduction, principles of quantum-scaled cement, principles of low-carbon cement, principles of fiber-reinforced concrete, principles of reactive rowder concrete, and principles of tailor-made recycled aggregate concrete. Reference articles on these topics in the extensive field of cementitious materials are incorporated into the book to demonstrate and discuss cement and concrete of composites. The book wants to promote best comprehension of building materials, bridge the gaps between engineering performances, science of materials, construction, in situ behaviors, and service/design life, encourage the developments of materials of low cost energy savings, and offer a platform for unconventional and unusual concretes. It is the intention of the book to have specific matters devoted to particular topic to concentrate attention on topical matters. The book will hopefully offer a unifying forum to bring together fabricators, designers, engineers, and scientists.

Within the above context, the book has several specific objectives. It wants to promote best comprehension of building materials, bridge the gaps between engineering performances, science of materials, construction, in situ behaviors, and service/design life, encourage the developments of materials of low cost energy savings, and offer a platform for unconventional and unusual concretes. Also, it is the objective of the book publish special matters dedicated to themes of emerging or current interests. The book aims to offer a unifying foundation for collaborations among fabricators, designers, engineers, and scientists of materials.

References

Aalbers TG, van der Sloot HA, Goumans JJJM (eds) (1994) Environmental aspects of construction with waste materials. Elsevier

Akcay B, Tasdemir MA (2010) Effects of distribution of lightweight aggregates on internal curing of concrete. Cem Concr Compos 32(8):611–616
Beushausen H, Gillmer M, Alexander M (2014) The influence of superabsorbent polymers on strength and durability properties of blended cement mortars. Cem Concr Compos 52:73–80
Chen Q et al. (2018) A stochastic micromechanical model for fiber-reinforced concrete using maximum entropy principle. Acta Mech 229(7):2719–2735
D'Alessandro A et al (2016) Investigations on scalable fabrication procedures for self-sensing carbon nanotube cement-matrix composites for SHM applications. Cem Concr Compos 65:200–213
Demis S, Efstathiou M, Papadakis V (2014) Computer-aided modeling of concrete service life. Cem Concr Compos 47:9–18
di Prisco M, Matteo C, Daniele D (2013) Fibre-reinforced concrete in fib model code 2010: principles, models and test validation. Struct Concr 14(4):342–361
Guo Z (2014) Principles of reinforced concrete. Butterworth-Heinemann
Hakim II, Putra N, Agustin PD (2020) Measurement of PCM-concrete composites thermal properties for energy conservation in building material. AIP Conf Proc. 2255(1)
Han B et al. (2014) Nanotip-induced ultrahigh pressure-sensitive composites: principles, properties and applications. Compos Part A Appl Sci Manuf 59:105–114
Han B et al. (2019) Nano-engineered cementitious composites: principles and practices. Springer
Harbulakova VO et al. (2014) Different aggressive media influence related to selected characteristics of concrete composites investigation. Int J Energy Environ Eng 5(2–3):82
Jones CA, Grasley ZC (2011) Short-term creep of cement paste during nanoindentation. Cem Concr Compos 33(1):12–18
Karakurt C, Kurama H, Topcu IB (2010) Utilization of natural zeolite in aerated concrete production. Cem Concr Compos 32(1):1–8
Kawashima S, Shah SP (2011) Early-age autogenous and drying shrinkage behavior of cellulose fiber-reinforced cementitious materials. Cem Concr Compos 33(2):201–208
Kene KS, Vairagade VS, Sathawane S (2012). Experimental study on behavior of steel and glass fiber reinforced concrete composites. Bonfring Int J Ind Eng Manage Sci 2(4):125–130
Klyuev SV et al (2018) The fiber-reinforced concrete constructions experimental research. Mater Sci Forum 931
Lai WL et al. (2010) Characterization of the deterioration of externally bonded CFRP-concrete composites using quantitative infrared thermography. Cem Concr Compos 32(9):740–746
Lamond J F, Pielert J H (2006). Significance of tests and properties of concrete and concrete-making materials. ASTM, West Conshohocken, PA
Li J, Yang E–H (2017) Macroscopic and microstructural properties of engineered cementitious composites incorporating recycled concrete fines. Cem Concr Compos 78:33–42
Lura P, Terrasi GP (2014) Reduction of fire spalling in high-performance concrete by means of superabsorbent polymers and polypropylene fibers: Small scale fire tests of carbon fiber reinforced plastic-prestressed self-compacting concrete. Cem Concr Compos 49:36–42
Makul N, Rattanadecho P, Agrawal DK (2014) Applications of microwave energy in cement and concrete—a review. Renew Sustain Energy Rev 37:715–733
Materazzi AL, Ubertini F, D'Alessandro A (2013) Carbon nanotube cement-based transducers for dynamic sensing of strain. Cem Concr Compos 37:2–11
Mo L, Deng M, Wang A (2012) Effects of MgO-based expansive additive on compensating the shrinkage of cement paste under non-wet curing conditions. Cem Concr Compos 34(3):377–383
Murtazaev SY et al. (2018) Impact of technogenic raw materials on the properties of high-quality concrete composites. In: International symposium engineering and earth sciences: applied and fundamental research (ISEES 2018). Atlantis Press
Novak J, Kohoutkova A (2018) Mechanical properties of concrete composites subject to elevated temperature. Fire Safety J 95:66–76
Ohama Y (2011) Concrete-polymer composites–the past, present and future. Key Eng Mater 466

Otero-Chans D et al. (2018) Experimental analysis of glued-in steel plates used as shear connectors in timber-concrete-composites. Eng Struct 170:1–10
Pan X et al. (2018) Effect of inorganic surface treatment on surface hardness and carbonation of cement-based materials. Cem Concr Compos 90:218–224
Pereira EB, Fischer G, Barros JA (2012) Effect of hybrid fiber reinforcement on the cracking process in fiber reinforced cementitious composites. Cem Concr Compos 34(10):1114–1123
Pourjavadi A et al (2013) Interactions between superabsorbent polymers and cement-based composites incorporating colloidal silica nanoparticles. Cem Concr Compos 37:196–204
Proske T et al (2018) Concretes made of efficient multi-composite cements with slag and limestone. Cem Concr Compos 89:107–119
Qiu Y et al (2013) Review on composite structural health monitoring based on fiber Bragg grating sensing principle. J Shanghai Jiaotong Univ (Sci) 18(2):129–139
Snoeck D, Pel L, De Belie N (2018) Superabsorbent polymers to mitigate plastic drying shrinkage in a cement paste as studied by NMR. Cem Concr Compos 93:54–62
Song Q et al (2018) Steel fibre content and interconnection induced electrochemical corrosion of ultra-high performance fibre reinforced concrete (UHPFRC). Cem Concr Compos 94:191–200
Szajerski P et al. (2020) Radiation induced strength enhancement of sulfur polymer concrete composites based on waste and residue fillers. J Clean Prod 271:122563
Ulm F-J et al (2010) Does microstructure matter for statistical nanoindentation techniques?. Cem Concr Compos 32(1):92–99
Wang X et al (2020) Interfacial characteristics of nano-engineered concrete composites. Constr Build Mater 259:119803
Wetzel A, Middendorf B (2019) Influence of silica fume on properties of fresh and hardened ultra-high performance concrete based on alkali-activated slag. Cem Concr Compos 100:53–59
Won M, Cho Y H, Tayabji S, Yuan J (2009) New technologies in construction and rehabilitation of Portland cement concrete pavement and bridge deck pavement. Am Soc Civ Eng
Wu C-R et al (2018) Improving the properties of recycled concrete aggregate with bio-deposition approach. Cem Concr Compos 94:248–254
Wyrzykowski M, Lura P (2013) Controlling the coefficient of thermal expansion of cementitious materials–a new application for superabsorbent polymers. Cem Concr Compos 35(1):49–58
Yang Z et al (2017) In-situ X-ray computed tomography characterisation of 3D fracture evolution and image-based numerical homogenisation of concrete. Cem Concr Compos 75:74–83
Zajac M et al (2018) Impact of microstructure on the performance of composite cements: why higher total porosity can result in higher strength. Cem Concr Compos 90:178–192

Chapter 2
Principles of Quantum-Scaled Cement

Abstract To begin with, the current study reviews the cement hydration state in the nanomaterial modeling presence. It is important to comprehend that nanotechnology is an active research part across the globe. In particular, the idea began after the carbon nanotube inventions, which is used in some areas that include machine components, electronic, and bio-mechanic. With that said, the advent of nanotechnology contributes to the development of materials which may be used to designs of high-performance concrete mixes. It is important to comprehend that nanosilicas react with calcium hydroxides to establish more of the strengths carrying the cement structures, calcium silicate hydrates. The paper in question indicates the development of correlations to differentiate the advantages when utilizing various sizes of nanosilica in cement paste. In fact, the mechanical characters of concrete substances count to a significant level on structural components as well as concepts that are adequate on micro-and nanoscales. The sizes of phases of the calcium silicate hydrates are the fundamental elements responsible for strengths and other features of cementitious substances, which depend on the ranges of few nanometers. Specifically, the C-S-H structures are like clays with thin solid layers separated gel pore with adsorbed and interlayer water.

Keywords Cement · Quantum-scale

2.1 Introduction

Cement mixes water and other components such as sand and ballast, to make a hard substance. Hydration is a chemical process that solidifies the cement and the other components to form a stone-like structure. The stone-like structure, concrete, is used to make pavements, pipe, and architectural structures. The high bonding structure enables the cement to be used in construction of motorways, bridges, fences, and poles. The structures made from the cement are long lasting due to their good quality. The structures are also able to accommodate heavy weight from machines such as vehicles. The fences and poles made from cement are not affected by pests. They are also waterproof. They give long-term services. Mr. John Smeaton, who is a British engineer, was the first person who made the concrete. The cement was referred

N. Makul, *Principles of Cement and Concrete Composites*, Structural Integrity 18,
https://doi.org/10.1007/978-3-030-69602-3_2

to as hydraulic cement due to its active nature to hold water. The engineer added pebbles that were used as a course aggregate. He then mixed the powdered brick not cement. Later, the competent engineer burnt ground limestone and clay together to form artificial cement. The burning removed the chemical properties of the course aggregate materials. The cement produced was stronger as compared to just crashing the limestone. The Portland cement which dominates concrete products was the engineer's effort. There are different strengths of cement, and the uses of cement differ according to its strength.

Hydraulic cement is used for general purposes where the particular attention for other cements is not required. The cement is used in making of concrete products. Hima cement is also a multi-purpose cement. The cement is used on construction sites to hold building materials such as bricks and blocks together. Beams and water retaining structures such as concrete tanks are also constructed using the Hima cement. Domestic poles and post-foundations are also made using Hima cement.

The needs to understand aspects influencing the concrete performance have sparked studies on basic constituents of concretes and cement who characteristics are determined by the calcium silicate hydrates (C-S-H), one of the most crucial cement hydration products (Garboczi 2009; Do 2013; Constantinides and Ulm 2004; Kuosa et al. 2014; Richardson 2004). Discussing the calcium silicate hydrates (C-S-H) nanostructures as well as descriptive models over the past two decades, 2000–2020, the present research describes the evolution of models characterized with the nanotechnology breakthrough and predictive and descriptive models. Along with the shortcomings of models, the outcomes and purposes of the recommended models are explained. The paper concluded that in essence, the contemporary simulations are still growing on the layered or colloidal models notwithstanding the huge potentials that advanced and nanotechnology modeling have provided in the field, instead of offering groundbreaking new approaches. Nevertheless, it is basically acknowledged that the shifts from descriptive to predictive models have been facilitated by the molecular and nanotechnology modeling. This saves resources and time by scientifically modifying the components of C-S-H and by inferring results of each extensive experiments such as sorption isotherm to derive various assemblages and structures that would have been impossible or hard in practice (Balaguru 2005; Zeman et al. 2010; Ulm et al. 2007; Sanchez and Sobolev 2010; Plassard et al. 2005; Powers 1900; Markeset and Myrdal 2009; Khalil 2016). In that sense, paired with analytical models, the new experimental models available are capable of providing justifications of advance cement nanoscience and the pioneering research, creating the ways to generation of nanomodified and innovative cements, with minimum contents of Portland cement.

The primary binding phase, the calcium silicate hydrates, and principal cement hydration product control the transport, mechanical, and chemical characteristics of hydrating cements (Garboczi 2009; Do 2013; Constantinides and Ulm 2004; Kuosa et al. 2014; Richardson 2004). It can be realized that cement science cannot move forward without quantitatively comprehending and mapping the C-S-H structure at the atom to the 100 nm levels, putting into account that the manner to improve the hydrated cement macroscopic mechanical characteristics (durability and strength as affected by aging, shrinkage, creeps, and porosities). The manipulations as well as observations of matters at the nanolevels can be accomplished with the applications of advanced analytical techniques and nanotechnology emergence, and these C-S-H nanostructure resolutions shall establish the grounds for further studies and evolutions of cementitious materials and advanced cements (Balaguru 2005; Zeman et al. 2010; Ulm et al. 2007; Sanchez and Sobolev 2010; Plassard et al. 2005; Powers 1900; Markeset and Myrdal 2009; Khalil 2016). These raise the questions as to whether nanotechnologies have succeeded to increase the hurdles set by the previous methods and by how far have the envelopes been improved in current studies.

First of all, it is evident that the nanostructures of calcium silicate hydrate gel play a vital role in numerous physical properties of concretes consisting of the significant engineering creep and shrinkage characteristics (Aalbers et al. 1994). In the modern world, a range of structural simulations and designs of the gel above as well as computational strategies for their validation have been established. In this case, the underlying C-S-H nanostructures are viewed as self-similar particles of agglomerations at two levels. In fact, computational approaches are presented for modeling images of transmission electron microscopies as well as computing sorption the design nanostructure features (Jennings et al. 2007). It is important to comprehend that nanoscale gel of C-S-H is the fundamental binding agents in concretes and cements, which are vital for the strengths and the long-term evolutions of the materials, as the amorphous textures of C-S-H mesoscales take a major part in material properties.

It is clear that C-S-H gels are the primary products of cement hydrations, and they contribute to concrete engineering characteristics (Ioannidou et al. 2016). Therefore, the C-S-H gel mechanical, physical, and microstructural mechanical characteristics present in cementation composites are bonded by the micro-poromechanical methods. It is noteworthy that the sophisticated physical and chemical structures of cement hydrate in concretes imply that challenges of underlying sciences still require being solved (Balaguru 2005). It is important to understand that a series of chemical responses occur during the hydration process of cement paste. For instance, once water is mixed with the cements, the tricalcium aluminates react with gypsum to yield ettringite and heat, and then, the tricalcium silicate is hydrated to generate the C-S-H, heat, as well as lime.

In the contemporary world, concrete researchers, as well as developers, take the edge of secondary cementation materials to offer concrete great strengths (Lee et al. 2014). In this case, utilizations of pozzolanic nanoparticles in the matrices of concretes are some of the most important approaches to break into the arenas of concrete designs. Pozzolanic nano-substances are used in the strength developments bearing cement paste crystals. In simple term, it is vital to understand that calcium silicate hydrates are important hydrated cement paste products and have sizes of particles in a range of 1–100 nm (Garboczi 2009). Thus, comprehending structures of nanoscale C-H-S would assist in improving the material performances. The new technologies precisely possess possibilities of modifying the hardened concrete properties and intricate hydration mechanisms of the cement-fine substances.

2.2 How Portland Cement is Made

It was in the year 1866, when David O. Saylor, Esaias Rehrig, and Adam Woolever started the Coplay Cement Company. Early in 1870s, Sailor began to experiment Portland cement. He used stones from his quarry. Sailor as lucky to discover that he could make Portland cement with the available materials (Aalbers et al. 1994). The Portland cement made from the crushed blocks and other components obtained was similar to that imported. However, the cement crumbled over a short period. Sailor realized that the proportions of raw materials used were not right and hence caused crumbling.

Portland cement is an artificial cement which was only available in England and Germany. In USA, only natural cement from natural rock was made (Lamond and Pielert 2006). Cement rock, coal, shale, and limestone were the raw materials used to manufacture Portland cement. Sailor used Lehigh deposits that were thought to be different from those employed in Europe. However, the deposits purified the cement made. Sailor discovered the Portland cement in 1875. The company received a gold medal for its high-quality product. The raw materials were obtained from a nearby quarry. Sailor had a challenge in ensuring correct proportions of the components. This is because, the components in the quarry contained different amounts in relation to the layers (Won et al. 2009). The chemists were highly involved to check out the correct proportion rates required for the manufacture of Portland cement.

In 1878, the company could produce 2500 barrels of Portland cement. The company had several kilns. In 1893, more furnaces were constructed for experimenting purposes. The kilns were only off in case of repair and maintenance purposes. With the addition of more furnaces, the company produced 500 barrels of Portland cement. It was a good improvement from 2500 barrels a month. Coplay Cement was taken over by the Essroc Company. Essroc Company is well known for the production of Saylor's Portland cement today.

Portland cement is produced by use of various raw materials. Their chemical combination carefully monitors silicon, aluminum, and iron. The end resulting product after the chemical process is cement. Limestone, chalk, and shells all combined with shale are heated to a very high temperature. The reaction is exothermic (Aalbers et al. 1994). The components form a rock-like substance that is grounded into a fine powder that is called cement. Portland cement passes through kiln. A kiln is a chamber that is subjected to very high temperatures. Physical and chemical combination of the raw materials takes place in the furnace.

Limestone and cement rock are mixed with water, crushed, and weighed separately. The weighing of the components is done by passing them through the kiln. Equal proportions of elements are ensured by the chemists in order to manufacture Portland cement that is free from crumbling a few days after its manufacture (Lamond and Pielert 2006). For the purpose of giving the mixture the required plasticity, the two rocks are combined with the natural cement. Lumps and balls are the most efficient form of the raw materials that enhance proper charging in the kilns. The plastic material formed is then made into balls and passed in the kiln for the chemical reaction. Balls increase the surface area to volume ratio and hence increasing the rate of reaction.

The cement kilns heat the ingredients at a very high temperature. The temperature used is about 2700 °F (Won et al. 2009). The materials that remain in the hot chamber combine chemically to form a substance called clinker. The clinker formed comes out of the kiln in the form of gray balls. The balls are small in the size if marble.

Clinker is then passed through the cooling chambers. The heat from the cooling chambers generated by the red-hot clinker is then returned back to the kiln. The furnace uses fuel as a source of energy. The recycling of heat saves energy and is more economical (Lamond and Pielert 2006). Chemists profoundly observe each and every stage in the manufacture of Portland cement. When the clinker is cooled, it is then mixed with limestone and gypsum. The product is then grinded into a fine powder. Finally, we have the Portland cement ready for transportation to the necessary construction sites.

2.3 Usage of Portland Cement

There are various uses of Portland cement that are associated with its competitive features. Portland cement is a high quality with a difference (Aalbers et al. 1994). As we have discussed earlier in the manufacturing process, each and every step is carefully observed to enhance quality final product.

For the benefit of its particle strength, Portland cement is used in the construction of pavements (Bye 1999). The cement can get strength very fast and hence reducing the curing period. It is used in construction of areas that need to be used quickly such as pavement.

Fencing poles are also made from the Portland cement. The durability of the cements leaves it at the advantage of constructing poles. The poles are also rust proof. Bridges are also built in the courtesy of Portland cement (Aalbers et al. 1994). The bridges hold a lot of weight from vehicles. The Portland cement is fit for the construction of bridges because of its high density. The bridges are, therefore, not at the risk of collapsing.

Water storage such as tanks and dam is constructed using Portland cement. The Portland cement can have a minimum heat produced during construction. The ability to bring the cooling effect makes the Portland cement suitable for the construction of dams, tanks and pipes (Lamond and Pielert 2006). Large massive building and structures are also built using Portland cement. The large building consists of thick walls that may crack in the case of scorching weather conditions. However, Portland cement can produce less heat at a slower rate and hence make it suitable to be used in large buildings foundation.

In areas where the structures are based on places with ground and soil water, Portland is used. A particular type of Portland moderate to sulfate reaction is used. The cement resists the sulfate from the groundwater from reacting with the construction materials (Won et al. 2009). The sulfate acts with the bricks and blocks hence making them weak over time. The building may collapse. Portland cement is there to serve all the unnecessary collision and to collapse of the structures.

Various complex reactions take place during cement hydration to yield amorphous, heterogeneous, complex multi-phase and porous cement paste matrices. Calcium silicate hydrate (C-S-H), the most significant hydration products of the cement-based substances are believed to be responsible for its cementitious characteristics like strength, hardness, and cohesion. Therefore, calcium silicate hydrate is considered as genomic codes cement-based substances. A detailed comprehension of the atomic structure or their chemistry might result in an enhanced control of the material characteristics at the nanoscales as show in Fig. 2.1. But continual transformations that take place in C-S-H atomic structures regarding environment and time make experimental investigations of their structure extremely hard. Nevertheless, C-S-H properties are already revealed. They are layered structures containing water molecule and calcium bonded between chains of tetrahedral silicate. Therefore, computational approaches are sought in investigating the C-S-H atomic structures that in turn shall assist in comprehending the associated experimental research. With the help of atomic configuration of some naturally occurring ideal crystals such as Jennite and Tobermorite that have close structural similarity with C-S-H, the C-S-H atomic is modeled. C-S-H mechanical characteristics such as shear modulus, bulk, Young's and Poisson forecasted with the assistance of energy optimization research are in close agreements with experimental data. On the bulk C-S-H characteristics, it is evident from the computation that the lengths of silicate chain have notable effects.

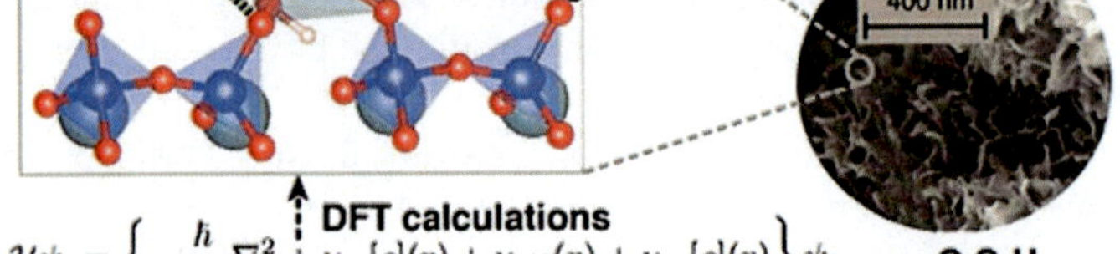

$$\mathcal{H}\psi_i = \left\{ -\frac{\hbar}{2m}\nabla^2 + \nu_{ee}[\rho](r) + \nu_{eN}(r) + \nu_{xc}[\rho](r) \right\} \psi_i$$

Fig. 2.1 Atomic-level structures of cementitious calcium silicate hydrates (Kumar et al. 2017)

2.4 Calcium Silicate Hydrate and Its Composites

Studies show that the primary cement products are amorphous materials referred to as the calcium silicate hydrates that form up to about 60% by paste volumes. In the chemistry of cements, C, S, and H denote calcium oxide (CaO), silicate (SiO_2), and water (H_2O), respectively. At times known as C-S-H gels, the available denotation of C-S-H demonstrates indefinite hydrates and stoichiometry (Sobolev and Gutiérrez 2005). In chemical responses of the silicate stages with water, investigations indicated that C-S-H is produced together with calcium hydroxides. The primary binding agents in the cementitious pastes are ideally C-S-H. It also obligated to its important properties. Thus, resolving the structures of preceding materials at nanoscales is viewed as significant parts of understanding and predicting their behaviors.

Researchers argued that C-H-S gels are multi-scale composites consisting of intermixtures, solids, and pores of C-H-S at the nanoindentation scales on C-S-H gels, and physical and mechanical gel characteristics can be affected by the volume fractions and porosities of the intermixtures (Kirkpatrick et al. 2005; El-Baky et al. 2013; Rahimi-Aghdam et al. 2017). It is crucial to comprehend how the heterogeneity created during the early hydration level influences the mechanical performances of hardened cement pastes and persists in C-S-H architectures. Thus, the bonds within cement hydrate can be groundbreaking. As well, controlling bonds in the cement hydrates can be important to smart mix designs of cementitious substances.

2.5 Formation and Properties

The molar ratios of calcium oxides to silicates in the materials are some of the important measures to define and manage C-S-H scheme characteristics in the calcium silicate hydrate formation (Yu and Lau 2015; Fonseca et al. 2011; Sorelli et al. 2008). From 1.2 to 2.1, the molar ratios of calcium oxides to silicates in the materials differ in phases of hydrated silicates. The mean ratio of molar ratios of calcium oxides to silicates in the materials is about 1.750. The schemes of C-S-H can be categorized into groups of low and high contents of limes. The calcium oxide/silicate ratio of about 1.10 in which chemical and physical characteristics conspicuously changes is applied to classify the C-S-H scheme. In addition, in C-S-H schemes, the water state is described where H_2O may be present in C-S-H interlayer architectures. In a nutshell, free water can be in the capillary pores between the clusters of C-S-H. Thus, water state distinctions are not just simple. The energies by which the particles of water are bonded in C-S-H differ over an extensive variety and can overlap to different areas. The actions existing between lime (calcium oxides) and silicas in excess water precisely contribute to the formations of schemes of jennite-like and tobermorite-like that are referred to as C-S-H (I) and (II). The hydrated C-S-H schemes explained above can be made by reacting calcium salts and sodium silicates in aqueous solutions. They are nevertheless less crystalline (Khan et al. 2017; Birgisson et al. 2012;

Smith and Virmani 2000). Pure substances in this stage are relevant for systematic study works on calcium silicate hydrates and are easy to produce.

2.6 Nanostructural Models of Calcium Silicate Hydrate

Different scholars have conducted extensive research on the nanostructure of calcium silicate hydrate; however, it is not clearly comprehended with proposed models arraying from layer-like to colloidal. Powers and Brownyard suggested one of the models, which defines C-S-H as colloidal materials. Specifically, the models illustrate that the gel components are bonded by van der Waal's bonds as well as the spaces existing among them are known as porosities of gels that water molecules can access. Later, Feldman and Sereda came up with a comprehensive model grounded on the extensive experimental research of hydrated cement schemes. In this design, the purpose of water is described in details, as well as the transformations in the mechanical features of C-S-H correlated to water contents are easily explained (Khan et al. 2017; Birgisson et al. 2012; Smith and Virmani 2000). The underlying property of the models above is interested in the structural natures of the materials in question. In particular, the structural purposes that are attributed to the interlayered water of calcium silicate hydrate depict irreversible character during the adsorption as well as desorption procedures.

The improvement in investigational strategies has contributed to the establishment of innovative simulations. For example, colloid designs of Jennings include bubbles of approximately 5 nm in diameters for calcium silicate hydrate as well as suggest the availability of intralobular pores as well as small pores of gels. The viabilities of utilizing structured designs for the materials in cementitious pastes are considered reasonable as for be the available modern work that employed helium inflow strategy as a nanostructure probe. In a nutshell, it is important to comprehend that the layered model is mismatched with the colloid design in its description of engineering and physic-chemical character.

2.7 Hydration Model for Cement Blended with Nanomaterials

It is important to understand that during the hydrations of cement pastes, the level of calcium hydroxide is proportional to the amount of hydration of cement. For instance, the degree of calcium hydroxide in cement-nanosilica blends during hydration can be identified with equation (Allen et al. 2007; Bentz 2007; Cerro et al. 2008; Jennings 2000). It can be assumed that the hydration of nanosilica consists of two processes including diffusion and phase-boundary process.

2.8 Improvement of Portland Cement Over Time

From the development of Portland cement Company, the quality of Portland cement has improved over time. Initially, during the manufacture of Portland cement, the cement could crumble after some time. The primary reason for crumbling was that, the proportions of the raw materials and components used were not proportional (Bye 1999). However, over time the default has been solved by involving chemists in each and every step of production to ensure that all the components are in good proportions. The Portland cements currently produced are strong, of high density, and good quality. It is free from crumbling.

The quantity of production of Portland cement has also improved with time. In 1875, the company could produce 2500 barrels of cement. After adding more kilns in 1900, the company produced 500 barrels of Portland cement in a day. The discovery of recycling heat energy used in cooling of clinkers played an imperative economic role (Aalbers et al. 1994). The company could produce more cements by sending back the heat obtained from cooling the clinker back to the kiln. Currently 5000 barrels of Portland cement are produced by the same company. With the use of development of technology, the production of Portland cement has become more efficient and effective.

For over a long period, only one standard Portland cement was produced. The cement was used for general purposes of construction. However, due to the dynamic demand in the construction field, the company came up with different types of Portland cement (Bye 1999). The various types of Portland cement are meant to meet different kinds of building needs. The development of different types of Portland cement marks its improvement with time. Storage and packaging of Portland cement has also improved over time. Initially, the packaging and weighing was done with manual labor and machines. Currently, with the use of technology, packaging and weighing is done by automatic packers and weighs. The method is fast. It saves time and also enhances accuracy (Aalbers et al. 1994). The packaging is done in bags of 42.5 kg that are ready for distribution and exportation.

2.9 Future Improvement and Importance of Portland Cement

Portland is a type of cement with a very high probability of success in the future. The Portland cement has already gained customer loyalty from the quality products. The price of Portland cement is also fair to its customer bearing in mind the quality of the product (Lamond and Pielert 2006). The marketing strategy used by Portland cement is inventive. The company does not add more product but improves the product already in the market.

The different types of Portland cement are also able to meet all the diversified needs of the customers. Therefore, in future, the Portland cement may become the domineering cement product in the markets. All the other brands may not be heard. If Portland continues with the same marketing strategy, all the customers from other cement brands may shift to Portland cement in future. The best aspect about Portland cement is that its product evolves with time. It grows to meet the changing needs and demands of the future.

The future importance of Portland cement is that it will be used as a reference. The structures built using Portland cement will be used as examples. The buildings that are currently built using Portland cement will still be active due to the competitive quality of Portland cement. Recommendations are made on which type of Portland cement to us in different sites (Won et al. 2009). For instance, the sulfate resistance used in sites with ground and oil water which produces sulfate elements. The various types recommended work efficiently when the engineers follow the instructions from Portland cement. Therefore, the future importance of Portland cement will be to clearly beyond reasonable doubt confirm the efficiency and effectiveness of the product from its past usage.

The Portland cement is artificially made. However, in the future, natural process should be considered. The natural process is challenging but has some added advantage as compared to the artificially made Portland cement (Lamond and Pielert 2006). The natural method uses less heat and is less polluted to the environment. The Portland should improve on preserving the environmental pollution by avoiding the artificial way. It is also very expensive. However, with the use of the natural method, less quality of Portland cement would be produced.

2.10 Conclusions

In conclusion, cement is very essential in the construction. The type of Portland cement to be used should be clearly defined. The high quality of Portland cement is as a result of the efficiency and accuracy observed during the manufacturing process. Chemists are there to ensure correct proportions of the raw materials components are used. The manufacturing process should be highly noted. It is very sensitive. Portland cement has improved over time by looking at customer needs and demands in the market. As a result, different types of Portland cement have been produced over time. Inventive strategy has also been highly employed in the manufacture of Portland cement. The approach has led to improvement in the quality of the Portland cement.

Overall, people need to comprehend that the emergence of nanotechnology usage in various areas, especially in concrete constructions, are regarded modern and numerous developments are still in the processes of commercializations. Therefore, enhancements in the comprehending of the properties of the hydration cement products at the nanoscale are required to enable the effective manipulations of the natures of cement-grounded substances. For instance, contemporary studies depict that the integration of nanosize calcium silicate hydrate atoms as seeds into cement schemes can be utilized to orient the compositions of the calcium silicate hydrate (C-S-H) as well as a monitor using modern established quantitative strategies.

References

Aalbers TG, van der Sloot HA, Goumans JJJM (eds) (1994) Environmental aspects of construction with waste materials. Elsevier

Allen AJ, Thomas JJ, Jennings HM (2007) Composition and density of nanoscale calcium-silicate-hydrate in cement. Nat Mater 6(4):311

Balaguru PN (2005) Nanotechnology and concrete: background, opportunities and challenges. In: Applications of nanotechnology in concrete design: proceedings of the international conference held at the University of Dundee, Scotland, UK on 7 July 2005. Thomas Telford Publishing, pp 113–122

Belkowitz JS (2009) An investigation of nano silica in the cement hydration process. Doctoral dissertation, University of Denver

Bentz DP (2007) Cement hydration: building bridges and dams at the microstructure level. Mater Struct 40(4):397–404

Bentz DP, Quenard DA, Baroghel-Bouny V, Garboczi EJ, Jennings, HM (1995) Modelling drying shrinkage of cement paste and mortar Part 1. Structural models from nanometres to millimetres. Mater Struct 28(8):450–458

Birgisson B, Mukhopadhyay AK, Geary G, Khan M, Sobolev K (2012) Nanotechnology in concrete materials: a synopsis. Transp Res E-Circular, (E-C170)

Bye GC (1999) Portland cement: composition, production and properties. Thomas Telford

Cerro Prada ME, Vázquez Gallo MJ, Alonso Trigueros JM, Romera Zarza AL (2008) Modelling hydration process of cement nanoparticles by using an agent-based molecular formation algorithm

Chandler MQ, Peters JF, Pelessone D (2012) Modeling nanomechanical behavior of calcium-silicate-hydrate (No. ERDC/GSL-TR-12-30). Engineer Research and Development Center Vicksburg MS Geotechnical and Structures Lab

Chandler M, Peters J, Pelessone D (2010) Modeling nanoindentation of calcium silicate hydrate. Transp Res Rec J Transp Res Board (2142):67–74

Constantinides G, Ulm FJ (2004) The effect of two types of CSH on the elasticity of cement-based materials: results from nanoindentation and micromechanical modeling. Cem Concr Res 34(1):67–80

Do QH (2013) Modelling properties of cement paste from microstructure

El-Baky SA, Yehia, S, Khalil IS (2013) Influence of nano-silica addition on properties of fresh and hardened cement mortar, vol 10. NANOCON Brno, Czech Republic EU, pp 16–18

Fonseca PC, Jennings HM, Andrade JE (2011) A nanoscale numerical model of calcium silicate hydrate. Mech Mater 43(8):408–419

Garboczi EJ (2009) Concrete nanoscience and nanotechnology: definitions and applications. Nanotechnol Constr 3:81–88

Grasberger S, Meschke G (2004) Thermo-hygro-mechanical degradation of concrete: from coupled 3D material modelling to durability-oriented multifield structural analyses. Mater Struct 37(4):244–256

Ioannidou K, Krakowiak KJ, Bauchy M, Hoover CG, Masoero E, Yip S, Ulm F-J, Levitz P, Pellenq R, Del Gado E (2016) Mesoscale texture of cement hydrates. Proc Nat Acad Sci 113(8):2029–2034

Jennings HM (2000) A model for the microstructure of calcium silicate hydrate in cement paste. Cem Concr Res 30(1):101–116

Jennings HM, Thomas JJ, Gevrenov JS, Constantinides G, Ulm FJ (2007) A multi-technique investigation of the nanoporosity of cement paste. Cem Concr Res 37(3):329–336

Khalil KA (2016) Effect of Nanosilica on the hydration characteristics and compressive strength of blended basalt cement pastes. Egy J Chem 59(4):573–595
Khan MU, Ahmad S, Al-Gahtani HJ (2017) Chloride-induced corrosion of steel in concrete: an overview on chloride diffusion and prediction of corrosion initiation time. Int J Corr
Kirkpatrick RJ, Kalinichev AG, Hou X, Struble L (2005) Experimental and molecular dynamics modeling studies of interlayer swelling: water incorporation in kanemite and ASR gel. Mater Struct 38(4):449–458
Kumar A, Walder BJ, Mohamed AK, Hofstetter A, Srinivasan B, Rossini AJ, Scrivener K, Emsley L, Bowen P (2017) The atomic-level structure of cementitious calcium silicate hydrate. J Phys Chem C 121(32):17188–17196
Kuosa H, Ferreira RM, Holt E, Leivo M, Vesikari E (2014) Effect of coupled deterioration by freeze–thaw, carbonation and chlorides on concrete service life. Cem Concr Compos 47:32–40
Lamond JF, Pielert JH (2006) Significance of tests and properties of concrete and concrete-making materials. Astm, West Conshohocken, PA
Lee HS, Cho HK, Wang XY (2014) Experimental investigation and theoretical modeling of nanosilica activity in concrete. J Nanomater 121
Markeset G, Myrdal R (2009) Modelling of reinforcement corrosion in concrete-State of the art. In: COIN P4 Operational service life design, SP 4.1 F Service life modelling and prediction
Montgomery J (2015) Effect of nano silica on the compressive strength of harden cement paste at different stages of hydration. Doctoral dissertation, North Carolina Agricultural and Technical State University
Murillo JSR, Mohamed A, Hodo W, Mohan RV, Rajendran A, Valisetty R (2016) Computational modeling of shear deformation and failure of nanoscale hydrated calcium silicate hydrate in cement paste: calcium silicate hydrate jennite. Int J Dam Mech 25(1):98–114
Plassard C, Lesniewska E, Pochard I, Nonat A (2005) Nanoscale experimental investigation of particle interactions at the origin of the cohesion of cement. Langmuir 21(16):7263–7270
Powers TC (1900) Physical properties of cement paste (154)
Rahimi-Aghdam S, Bažant ZP, Qomi MA, (2017) Cement hydration from hours to centuries controlled by diffusion through barrier shells of CSH. J Mech Phys Solids 99:211–224
Raki L, Beaudoin J, Alizadeh R, Makar J, Sato T (2010) Cement and concrete nanoscience and nanotechnology. Materials 3(2):918–942
Richardson IG (2004) Tobermorite/jennite-and tobermorite/calcium hydroxide-based models for the structure of CSH: applicability to hardened pastes of tricalcium silicate, β-dicalcium silicate, Portland cement, and blends of Portland cement with blast-furnace slag, metakaolin, or silica fume. Cem Concr Res 34(9):1733–1777
Sanchez F, Sobolev K (2010) Nanotechnology in concrete—a review. Constr Build Mater 24(11):2060–2071
Smith JL, Virmani YP (2000) Materials and methods for corrosion control of reinforced and prestressed concrete structures in new construction (No. FHWA-RD-00-081)
Sobolev K, Gutiérrez MF, (2005) How nanotechnology can change the concrete world. Am Ceramic Soc Bull 84(10):14
Sorelli L, Constantinides G, Ulm FJ, Toutlemonde F (2008) The nano-mechanical signature of ultra high performance concrete by statistical nanoindentation techniques. Cem Concr Res 38(12):1447–1456
Ulm FJ, Vandamme M, Bobko C, Alberto Ortega J, Tai K, Ortiz C (2007) Statistical indentation techniques for hydrated nanocomposites: concrete, bone, and shale. J Am Ceramic Soc 90(9):2677–2692
Won M, Cho YH, Tayabji S, Yuan J (2009) New technologies in construction and rehabilitation of Portland cement concrete pavement and bridge deck pavement. Am Soc Civil Eng
Yu R, Spiesz P, Brouwers HJH (2014) Effect of nano-silica on the hydration and microstructure development of ultra-high performance concrete (UHPC) with a low binder amount. Constr Build Mater 65:140–150
Yu Z, Lau D (2015) Nano-and mesoscale modeling of cement matrix. Nanoscale Res Lett 10(1):173

Zandi HK (2010) Structural behaviour of deteriorated concrete structures. Chalmers University of Technology
Zeman JC, Edwards JR, Lange DA, Barkan CP (2010) Investigation of potential concrete tie rail seat deterioration mechanisms: cavitation erosion and hydraulic pressure cracking. In: Proceedings of the transportation research board 89th annual meeting, Washington, DC

Chapter 3
Principles of Low-Carbon Cement

Abstract The cement industry constitutes a severe threat to ecology, including through its negative impact on the climate, due to the high level of carbon dioxide (CO_2) emissions associated with it. Given that this is the case, the modern world is looking for alternatives in order to preserve the environment for future generations. The eventual goal, therefore, is for industry to stop emitting carbon into the air. Many effective steps can be taken by industry leaders to achieve lower carbon emission targets to improve local ecological systems. This paper discusses the ways in which CO_2 is measured and alternatives to the standard methods through which hydraulic cement is produced in order to reduce CO_2 emissions. The benefits of using alternative methods, specifically relying on kilns and/or synthetic fuels, are identified and discussed. An assessment of the conditions needed for the industrial production of new cementitious systems in which clinker-calcined limestone and low-carbon clay are used is also presented. Additionally, an account of the clinkerization process of low-carbon cement (LCC) is provided. The new materials are shown to meet global standards in applications such as the production of hollow concrete blocks and precast concrete. In a comparison between Portland cement and the new materials, no major differences were found in either the mechanical or rheological features. An environmental ternary cement assessment is also reported that includes comparisons with other industrially blended cements. LCCs are shown as having the ability to reduce carbon emissions from cement production by more than 30%.

Keywords Low carbon cement · Cement production · Ecological impact

3.1 Introduction

Four main processes in the production of cement have a great impact on the resulting cement qualities, its fuel consumption, and the kind and amount of pollutants generated (Imbabi et al. 2012): the preheating of raw materials, calcination procedures, the combustion of clinker, and the cooling of clinker (Moya et al. 2011). The process through which cement is manufactured is a very rigorous one in which the raw materials used and the processes followed are responsible for the emission of carbon. However, many effective measures can be applied to the production of cement in

N. Makul, *Principles of Cement and Concrete Composites*, Structural Integrity 18,
https://doi.org/10.1007/978-3-030-69602-3_3

order to reduce the emission of carbon dioxide (CO_2). Such measures have the benefit of minimizing the industry's negative impact on the domestic ecological system and of improving the effectiveness of cement manufacturing (Phair 2006).

As calcination procedures have a significant negative impact on the environment, it is advisable to replace the raw materials currently used in most cement with alternatives that have no carbonates in their cement composition and structure Miculčić et al. 2013). The quantity of "pure" cement in a given cement mixture can be decreased by replacing some of the standard materials with pozzolanic materials with comparable ability to act as a cement in relation to binding all the products in the mix. Specifically, industrial waste products such as fly ash, slag, silica fume, and rice husk ash, which are ordinarily used as landfill, can act as a pozzolana product (U.S. 2018). The government along with world and local industrialists needs to collaborate on finding ways to reduce the impact of carbon on the environment and to make agreements to set and adhere to new standards on this point.

An estimated 17–25 gigatons of concrete are produced each year, making cement one of the most used materials on earth (Smith et al. 2002). The manufacturing of Portland cement accounts for over 5% or 2.4 gigatons of emissions of global anthropogenic CO_2 every year, according to research published in 2010 by the U.S. Energy Department. Each ton of Portland cement clinker generated translates to about 810 kg of CO_2 emissions (Ludwig and Zhang 2015). The cement industry, which participates in the World Business Council for Sustainable Developments' Cement Sustainability Initiatives, has acknowledged the need to reduce both its energy and CO_2 footprints. The Cement Sustainability Initiatives, which have driven a comprehensive strategy to reduce CO_2 emissions through, for example, the use of sequestration technologies, alternative fuels, and supplementary cementitious materials, have made notable progress (World Business Council for Sustainable Development 2002). There is, however, much more work to be done.

In the endeavor to minimize the CO_2 footprint of ordinary Portland cement, several new chemical compositions for cement products have been introduced in recent years (Mehta and Monteire 2006). These new types of cement have originated from both within and outside the traditional cement industry. They involve products such as the following:

- Lafarge produces Aether cement. It is principally composed of calcium sulfoaluminates ($4CaO \cdot 3Al_2O_3 \cdot SiO_2$) and belites ($2CaO \cdot SiO_2$).
- Celitement GmbH manufactures Celitement cement, which includes calcium hydro-silicates in its composition.
- Novacem produces cement that includes magnesium oxide (MgO) obtained from naturally occurring magnesium silicate.
- The Zeobond Pty Company makes E-Crete, a concrete product that includes geopolymers extracted from fly ash activation.

However, these products' penetration into and acceptance in concrete markets has been slow, although they all have positive characteristics (Hasanbeigi et al. 2010). As well as achieving greater acceptance of comparatively sustainable cement products, multiple steps must be taken to further "green" the operations of the cement and concrete sectors. The principal steps in this process are as follows:

- Modify standard concrete manufacturing procedures in order to minimize greenhouse gas emissions and the production of toxic substances and gases such as CO_2.
- Provide access to productive and viable CO_2-sequestration technologies to further reduce the industry's undesirable ecological impact.
- Demonstrate that enhanced cement end products can serve as alternatives to traditional Portland cement.
- Achieve the gains stated above while remaining compatible with the industry's existing operations and infrastructure.

In addressing the negative environmental impact of cement production, Solidia cement provides an innovative approach to minimizing emissions of CO_2 originating from the manufacturing and applications of cement products (Gartner 2004). In 2019, more than six billion tons of cement were produced. About twenty billion cubic meters (m^3) of cement were produced in that one year given that the average cement content per cubic meter (m^3) of concrete is around 250 kg. This is similar to about 2.0 m^3 of concrete for every person on earth. No other material is generated in such huge amounts. About 4–10% of anthropogenic aggregate CO_2 emissions are associated with the production of concrete given the enormous amount of concrete generated and used in the construction and building industry. About 80% of concrete emissions are related to the manufacture of cement (Turner and Collins 2013). Further, cement production accounts for about 4–6% of CO_2 emissions worldwide. Any improvement in the production of cement has the potential to significantly reduce the amount of CO_2 emitted into the atmosphere.

By 2050, it is forecast that the demand for cement worldwide will double, with most of the increase coming from developing nations where infrastructure has yet to be developed. Further, population increases will put pressure on the growth of existing urban centers. The processes of cement production include combustion of the raw material at an extremely high temperature to generate clinker (Rashad and Zeedan 2011). The consumption of energy is between 3.0 and 6.5 GJ/tons of clinker depending on how the raw material is processed. In cement production, greenhouse gases such as CO_2 are emitted. Calcium carbonate (CaO) accounts for about 70–80% of the raw materials used to manufacture clinker (Gartner 2004). Therefore, CaO is the major component of cement. For each ton of calcined materials, calcium carbonate decomposes into calcium oxide and emits 0.44 tons of CO_2 into the atmosphere at around 900 °C. The rest of the carbon released originates from electricity utilized for grindings and fuels in the firing process (Rashad 2015). Cement production emission rates range between 0.66 and 1.10 tons CO_2/ton cement depending on the efficiency of the kilns used.

One beneficial approach to realizing significant reductions in the release of CO_2 during the production of cement is that of minimizing the quantity of clinker (Rashad et al. 2013). This is the case because the production of clinker is the main cause of CO_2

emissions. A reduction in clinker can be achieved by using supplementary cementitious concretes as a substitute (Fodor and Klemeš 2012). The rate of substitution is about 30% for most cement plants; however, according to the last study in which this was measured (Lund 2007), this is not sufficient to radically reduce the amount of carbon released from cement production internationally.

For this paper, the authors have designed new cementitious systems that increase the extent to which alternative materials that lead to lower CO_2 emissions are substituted for clinker to 50% without greatly affecting the performance of the resulting cement as compared to that of ordinary Portland cement (World Business Council for Sustainable Development 2009). The new model is founded on synergies between the carbonate from limestone and aluminate supplied by calcined kaolinite clays, which allow for a greater rate of substitution and improve the pozzolanic reactions of the calcine clays (Deja et al. 2010). The CO_2 emissions related to clinker are reduced significantly in the new model. The calcium carbonates added to the system are not calcined, and therefore, no additional CO_2 is emitted into the atmosphere (Kartini et al. 2014). This validates the tag "low-carbon cement (LCC)" given to the cement described.

Limestone and calcined clays, the two supplementary materials used in concretes have greater availability than do other supplementary cementitious concretes. Currently used by the construction sector, low-grade kaolinite clays have proven to be appropriate alternatives to pure kaolinite metakaolin (Yang et al. 2016). As compared to deposits of pure kaolinite clay, reserves of low-grade kaolinite are huge and more broadly geographically distributed. In the same order of magnitude as the present production of cement worldwide, industrial trials conducted at cement plants in various parts of the world for new cement manufacture refer to a large scale of production (Crossin 2015).

3.2 Fundamentals of LCC

To date, it has not been found economically feasible to produce cement that matches the quality of Portland cement using any mineral as a substitute material (European commission 2018). Researchers have investigated the impact of minerals used as substitutes on the mechanical features of cement and the emission of CO_2. According to these research studies by decreasing the proportion of clinker by partially replacing it with other materials, predominantly waste materials, and reducing the electrical and thermal energy consumption, it is actually relatively easy to decrease CO_2 emissions. Because of the presence of a large amount of CO_2 material in fuel gaseous substances, an efficient way to decrease CO_2 emissions from the manufacture of cement is to absorb carbon from the gases and consequent to store it (Li et al. 2010). Researchers evaluated one of the newly built industries in London: They analyzed the technology used to capture CO_2 and the amount captured and identified the obstacles to capturing CO_2 that emerged. They evaluated calcium loop and amine scrubbing technology as prospective carbon capture and storage (CCS) tools in the cement sector (Bosoaga et al. 2009). The results showed that calcium looping can be used to remove lime and, therefore, could be used to produce clinker and so reduce the emission of CO_2. Moreover, it is also possible to reduce CO_2 emissions by improving the process through which clinker is manufactured. The best technology is the use of kilns along with preheating multi-phase cyclones. Cao et al. (2016) investigated the energy that contributed to the use of technology to reduce carbon emissions. They used life cycle assessment techniques to distinguish between the carbon emissions of the new technology and those of the prior clinker-producing plant. The researchers offered the probable of computational fluid dynamic (CFD) to optimize the cement calciner functional conditions and to assist in reducing CO_2 emissions arising from the production of cement. It was found that replacing greenhouse fuels can play a major role in reducing CO_2 emission (Pacheco-Torgal et al. 2008). The researchers offered waste products as a substitute for clinker and identified the positive pros and cons of existing waste-to-energy techniques.

3.3 Chemical Mechanisms for the Formation of Low-Carbon Cement (LCC)

Cement is manufactured according to a three-phase process: the preparation of raw materials, the production of clinker, and the grinding of clinker with other constituents (Fig. 3.1). Various raw materials are grinded and mixed into homogenous powders, from which clinker is generated in a high-temperature kiln, thereby giving rise to the emission of CO_2.

Solidia cement is chiefly composed of lime-containing silicate phases such as rankinites ($3CaO \cdot 2SiO_2$) and pseudowollastonites or wollastonites ($CaO \cdot SiO_2$). The composition of Solidia cement clinker includes 42–48 wt% CaO (Ojan et al. 2016). This composition is quite unlike that of ordinary Portland cement, which consists of CaO-rich phases including tetra-calcium aluminoferrites ($4CaO \cdot Al_2O_3 \cdot Fe_2O_3$) and tricalcium aluminates ($3CaO \cdot Al_2O_3$). Typically, ordinary Portland cement clinker is 65–70% CaO.

The differences and similarities between the chemical composition of ordinary Portland cement and that of Solidia cement are noteworthy (Italcementi Group 2012). Ordinary Portland and Solidia cements are similar inasmuch as they are manufactured with the same unit operations, from the same raw materials, and in the same production plants (Bosoaga et al. 2009). For more efficient and faster implementation of new products, it is essential that new technologies are compatible with the cement industry's existing infrastructure. The chemical composition of Solidia cement differs from that of Portland cement (Scrivener 2016) and as such leads to a reduction in the emission of CO_2 related to cement production and provides a foundation for the sequestration of CO_2 during cement curing.

For both Portland cement and Solidia cement, production begins by establishing limestone mixtures as a source of CaO and shale, sand, or clay as a source of silicon

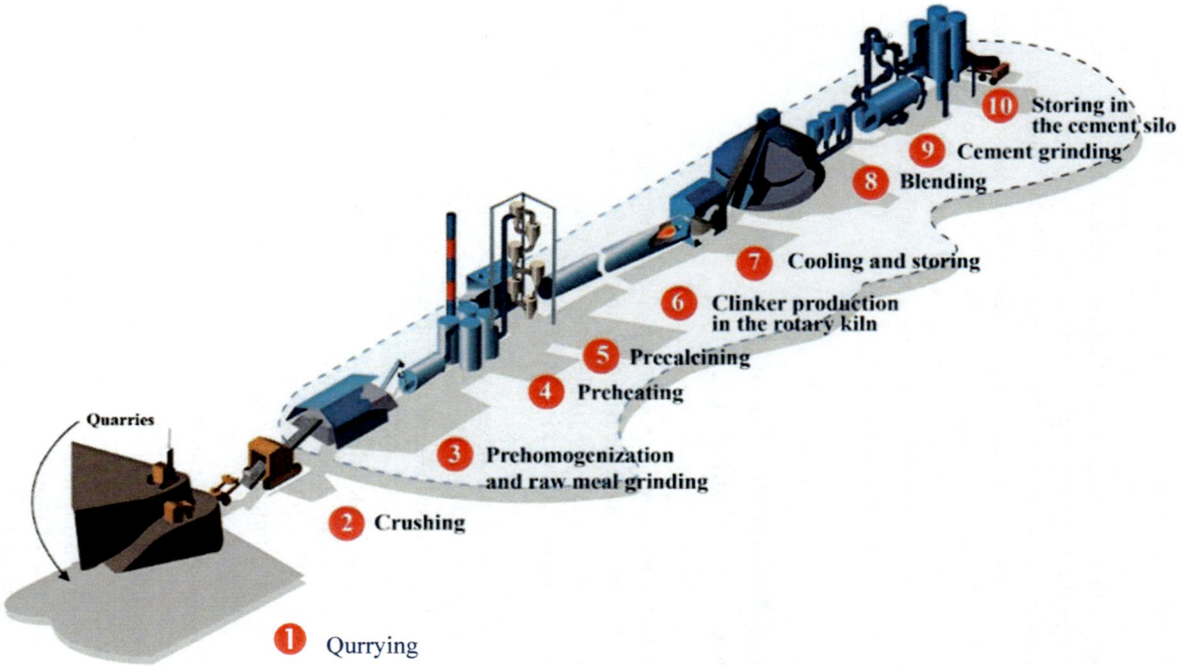

Fig. 3.1 Production and grinding of clinkers with other constituents to generate cement (Damtoft et al. 2006)

dioxide (SiO_2) (Georgiopoulou and Lyberatos 2018). These materials are available in almost every cement factory and from quarries worldwide. Typically, more than 70% of the composition of lime-rich Portland cement is calcium carbonate ($CaCO_3$) (Bontempi 2017). For low-lime Solidia cement, this figure drops to just about 50% $CaCO_3$. These differences offer a clear opportunity to minimize the cement clinker CO_2 footprint of the cement industry. The first significant reaction starts at around 800 °C as the raw materials are heated in cement production operations (Galvez-Martos and Schoenberger 2014). The limestone calcines or decomposes to form gaseous CO_2 and CaO at these temperatures according to Reaction (3.1):

$$CaCO_3(S)gCaO(S) + CO_2(g) \tag{3.1}$$

The calcination reactions release around 540 kg of CO_2/ton of Portland cement clinker generated with mixtures of raw materials for the lime-rich chemical composition of ordinary Portland cement (Bontempi 2017). Representing a 30% reduction, the mixtures of low-lime raw materials utilized to produce Solidia cement emit about 375 kg of CO_2/ton of Solidia cement clinker.

In cement production, the next important chemical reaction occurs at a temperature at which the raw materials partly fuse, sinter, and react to form clinker nodules (Bosoaga et al. 2009). This reaction takes place at about 1450 °C for Portland cement and leads to the formation of tetra-calcium aluminoferrite, belite, and alite compounds. The reactions through which these compounds form are represented by Reactions (3.2)–(3.5):

$$\begin{aligned} &4CaO(s) + Al_2O_3(s) + Fe_2O_3(s) \rightarrow 4CaO \cdot Al_2O_3 \cdot Fe_2O_3(s) \\ &\quad \text{(tetra - calcium aluminoferrites)} \end{aligned} \tag{3.2}$$

$$3CaO(s) + Al_2O_3(s) \rightarrow 3CaO \cdot Al_2O_3(s)\text{(tricalcium aluminates)} \tag{3.3}$$

$$2CaO(s) + SiO_2(s) \rightarrow 2CaO \cdot SiO_2(s)\text{(belites)} \tag{3.4}$$

$$3CaO(s) + SiO_2(s) \rightarrow 3CaO \cdot SiO_2(s)\text{(alites)} \tag{3.5}$$

At about 1200 °C, for low-lime Solidia cement, the chemical composition allows the raw materials to react, fuse, and sinter to form clinker (Gartner and MacPhee 2011). The resulting rankinites and pseudowollastonite or wollastonite phases for Solidia cement clinker take place according to Reactions (3.6) and (3.7):

$$CaO(s) + SiO_2(s) \rightarrow CaO \cdot SiO_2(s)\text{(wollastonites/pseudowollastonites)} \tag{3.6}$$

$$3CaO(s) + 2SiO_2(s) \rightarrow 3CaO \cdot 2SiO_2(s)\text{(rankinites)} \tag{3.7}$$

The ability to manufacture Solidia cement clinker at lower peak temperatures than is needed for ordinary Portland cement directly translates into decreased fossil fuel consumption. At a peak temperature of 1450 °C, the fuel combustion required to generate one ton of ordinary Portland cement clinker generates about 270 kg of CO_2 (Damtoft et al. 2006). At 1200 °C, one ton of Solidia cement clinker emits as little as 190 kg of CO_2 from fuel oxidation. As compared to the production of ordinary Portland cement, Solidia cement can be produced with CO_2 emissions of up to 30% less.

In cement manufacturing, pozzolan is widely utilized as supplementary cementitious concrete to replace clinker (Van Den Heede and De Belie 2012). For a silica-rich pozzolanic system, pozzolan reacts as expressed by Reaction (3.8):

$$SiO_2(s) + 1.5Ca(OH)_2 + 1.8H_2O \rightarrow 1.5CaO \cdot SiO_2 \cdot 2.8H_2O \tag{3.8}$$

The amount of clinker used to replace pozzolan depends on the amount of calcium hydroxide ($Ca(OH)_2$) generated during the hydration of the cement (Stafford et al. 2016). The amount of calcium hydroxide generated during hydration is not sufficient to react with the pozzolanic materials for substitution ratios above 15 wt% or in a pure pozzolanic system, as shown as Table 3.1. Therefore, the unreacted materials perform like fillers. As pozzolanic reaction kinetics are often slower than in pure systems of clinker, pozzolan used to replace clinker at a high level may compromise the binder mechanical characteristics (Mindress et al. 2003). The majority of cement manufacturers restrict the materials used to substitute clinker to no more than 24–36 wt% for this reason.

Metakaolin ($Al_2Si_2O_7$) is a very reactive pozzolan generated via the calcination of clay rich in kaolinite minerals. Metakaolin reactivity depends on various factors such as the temperature and rate of cooling and heating in given regimens (Valderrama et al. 2012). Even though de-hydroxylation of the clay is known to start at 500 °C, the optimum calcination temperature is 700–800 °C. Reaction (3.9) describes the metakaolin pozzolanic reactions with $Ca(OH)_2$ pastes:

Table 3.1 Relationships between availability of Portlandites and clinker substitutions in the blended cements (Mindess et al. 2003)

Compound	Cement 0 wt% substitutions (g)	Cement 15 wt% substitutions (g)	Cement 20 wt% substitutions (g)	Cement 25 wt% substitutions (g)
Available calcium hydroxide	27	23	21.6	20.3
Pozzolans	0	15[a]	20.0[a]	25.0[a]
Calcium hydroxide needed for a complete reaction[a]	0	19.5	25.9	32.4

[a]Computed by the authors (Mindess et al. 2003) putting into consideration that 70% of pozzolans react

$$Al_2Si_2O_7 + 5Ca(OH)_2 + 3H_2O \rightarrow 4CaO \cdot Al_2O_3 \cdot 13H_2O + 2CaO \cdot SiO_2 \cdot H_2O \quad (3.9)$$

The use of metakaolin as a supplementary cementitious material the volumes of new alumina stages in the systems because its composition to be alumina-rich compound (Josa et al. 2007). The alumina phases with calcium carbonate form hemi-carboaluminate ($3CaO \cdot Al_2O_3 \cdot 0.5Ca(OH)_2 \cdot 0.5CaCO_3 \cdot 11.5H_2O$) and moncar-boaluminate ($3CaO \cdot Al_2O_3 \cdot CaCO_3 \cdot 11H_2O$) if $CaCO_3$ and are introduced to the pore cement matrix systems through external sources. It is possible to substitute masses of clinker on the basis of these principles by equivalent masses of metakaolin and calcium carbonate reacted in a ratio of 2:1 to form hydration compounds capable of filling the pore cement matrix systems and contributing to the strength of cement produced (Von Bahr et al. 2003). Experimental lab results indicate that it is possible to substitute up to 45 wt% of clinker without compromising compressive strength (8). Specifically, 90% of the strength of ordinary Portland cements can be obtained with materials used to replace 60% of the clinker usually used.

3.4 Production of LCC

Several steps are taken to reduce CO_2 emissions as per Kania's model (Boesch and Hellweg 2010). Kania suggested reducing clinker to the ratio of cement by incorporating various additives. This ratio is now low at around 0.57; however, there is no guarantee that it can be reduced further if the performance and reliability characteristics of ordinary Portland cement are to be maintained. Based on the model, Kania also suggested using fossil fuels as a means of reducing the emission of CO_2. Another benefit is the decrease in the energy costs associated with the production of cement, which relies on an environmentally friendly technique of utilizing waste to improve the energy efficiency of the kilns used in cement production, which is another way to reduce carbon emission (Gäbel et al. 2004). Lastly, co-producing synthetic fuel is a way to use renewable energy in manufacturing cement procedures, and recycling the CO_2 from gases will reduce the emission of CO_2. Steps in this direction are driven by environmental strategy and the legal and regulatory environment and can be incorporated into cement production only if shown to be cost-effective.

Countries are increasingly making plans and adopting measures to reduce the emission of CO_2 in an effort to address climate change and are engaging in monitoring and research relating to climate change likewise (ISO 2006). The effectiveness of the energy of the cement industry is analyzed in such a way that the energy consumption of any specific cement plant is associated with the energy usage of the benchmark (Hendriks et al. 2004). This energy consumption value is used in analyzing the effectiveness of measures put in place to improve the processes involved in the production of cement. The consumption of thermal energy of a kiln procedure is 3.7 GJ/ton clinker (ISO 2006). There is still much room to improve energy efficiency such that steps should be taken to further develop efficiency by adding calcine before the materials are placed in the kiln. One of the European Union's (EU) most important projects is to reduce the emission of CO_2 from the cement industry, and the EU is, in fact, working to find alternatives to current production methods for ecological reasons (Pennington et al. 2004). The study of Josa et al. (2004) evaluated the ecological effectiveness of three CO_2 emission steps that have been taken in some parts of the industry and prioritized them according to the extent to which each reduced CO_2 emissions. The steps taken were those of using substitute fuels, using a more effective kiln procedure, and producing synthetic energy. The results show that around 0.2 tons of carbon could be prevented, accounting for 40% of CO_2 emissions. Based on these results, steps already identified for reducing CO_2 emissions can be expected to prove effective in meeting targets and producing cement in a more sustainable way.

The production of cement and concrete is among the most significant sources of CO_2 emissions, accounting for 4–8% of total emissions of anthropogenic CO_2. Because more than 50% of CO_2 emissions from cement production are process-associated, decarbonizing cement and concrete manufacturing is a difficult proposition. Nevertheless, new technologies and materials are being examined to minimize the CO_2 footprint of this sector (Flower and Sanjayan 2007). These include the production and application of alternative cements, mortars, and concretes, and reclaimed materials with lower emissions of CO_2, as well as materials and technologies for the sequestration of carbon (Rebitzer et al. 2004). In the coming decades, it is nevertheless still predicted that traditional ordinary Portland-based cement will continue to dominate construction applications. This study topic concentrates on present developments in technologies and materials that result in concrete, cement, and mortar production with minimized emission of greenhouse gases as compared to that of traditional ordinary Portland cement (Chen et al. 2010). In this regard, the purpose of the study (Chen et al. 2010) is to promote the sustainable design and construction of structural materials and built environments. Considerations that fall within the topic scope include the following:

- Replacements such as geo-polymers and alkali-activated materials to replace the binders used in ordinary Portland cement in order to develop alternative concrete and cement.
- Hybrid alkali-activated materials and supplementary concrete materials used to partially replace ordinary Portland cement in concrete and mortar.
- Technologies to recycle or reuse concrete structures and materials.
- Aggregates, mortars, and concretes produced with reclaimed materials.
- Technologies for the sequestration of carbon in the production of binder, clinker, and cement, for example, CO_2 curing.
- Materials and technologies for the sequestration of carbon in concrete and mortar such as biochar.

Efforts are being made to meet the goals of the United Nations (UN) focused on sustainable development at a time when climate change has drawn international attention (ATILH 2002). However, the most widely utilized man-made material on earth and an important source of emissions of CO_2, concrete is often overlooked in conversations about climate change (Stemmermann et al. 2010). The production of cement, a major ingredient in concrete, accounts for 7% of emissions of CO_2 globally—more than any industry except for the steel and iron industry. The processes of cement production are responsible for 95% of the carbon footprint of concrete (Stemmermann et al. 2011). To meet the sustainable development targets (referred to as two-degree scenario (2DS) targets) of the International Energy Agency, cement manufacturers required to reduce their carbon output at a yearly rate of 0.3% per ton of cement until 2030 (Van et al. 2003). Cement factories that fail to improve their energy efficiency and fail to adopt low-carbon processes might face penalties and fines for non-adherence.

LCCs are manufactured by utilizing the limestone and clinker-calcined clay ternary systems explained above. In these systems, medium-purity kaolinite clays have proven to be the best alternatives to metakaolin in terms both of being readily available and of minimizing the manufacturing costs of cement (Gartner 2004). With the use of metakaolin, no additional CO_2 is released into the atmosphere because the calcium carbonates introduced to the systems are not calcined.

The literature already includes studies focused on ternary Portland limestone blends with alumina pozzolans. De Weert et al. (2001) and Moensgaard et al. (2011) have researched systems with fly ash, limestone, and/or ordinary Portland cement and reported on the efficacy of these materials in the manufacture of LCC. The research results showed improvements to the mechanical characteristics caused by synergies between the limestone and fly ash that favor carbo-aluminate phase formation and thus make a positive contribution to system performance. Similar systems and the synergies established between limestone and metakaolin have been studied by Antoni et al. (2012). Their contributions to improving the cementitious system's mechanical properties when used to replace large amounts of clinker have been verified.

3.4.1 What Drives High Carbon Output?

In the manufacturing process, clinker formation accounts for 60–70% of the total emissions of CO_2. The remaining 30–40% of CO_2 emissions originate from the fuel combustion that powers the manufacturing process, as shown in Fig. 3.2 (Huntzinger

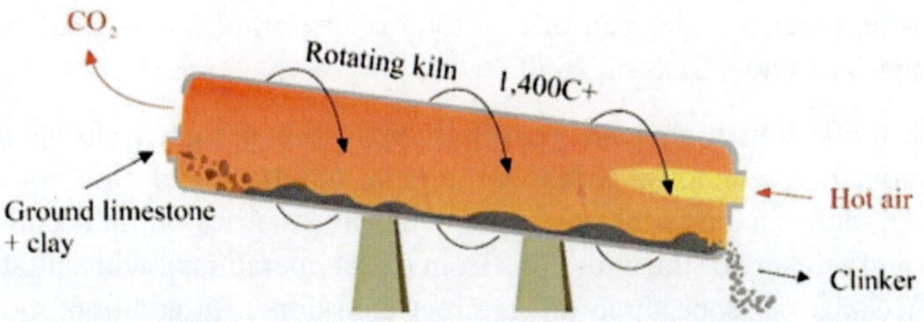

Fig. 3.2 Combustion rate and moisture contents of raw materials (Huntzinger and Eatmon 2009)

and Eatmon 2009). In addition, factors such as the availability of alternative fuels, the combustion rate and moisture content of the raw material, and the capacity of the cement factory also influence the carbon footprint and energy efficiency of cement production. In the manufacturing process of cement, the formation stage of the clinker cannot be replaced because there are no practical alternative materials presently available to replace limestone.

3.4.2 Opportunities for Carbon Reduction

Studies suggest that combinations of various initiatives can minimize CO_2 emissions across the value chain of the industry. However, there is no single all-encompassing approach to achieving carbon-neutrality within the industry:

- Energy efficiency: Compared to performing calcination in state-of-the-art process kilns, the production of clinker in wet kilns utilizes around 85% more energy (Febelcem 2006). In the best-case scenario, with the substitution of all wet kilns with technologically advanced dry kilns, an improvement of 10% could be realized internationally by 2050.
- Alternative fuels: Alternative fuels could comprise waste materials such as biomass, sewage sludge, and waste oils. The clinker firing is conventionally carried out with fuel oil, coal, or natural gases (Habert et al. 2010). A cumulative reduction in CO_2 emissions of 12% could be achieved internationally under the 2DS by 2050 by replacing fossil fuels with these alternative fuels.
- Minimizing clinker factors: The clinker factor is the average clinker to cement ratio, and the 2014 global ratio of the average clinker of 0.64 could be reduced to 0.60 by 2050 (Damtoft et al. 2008). Such a reduction could be achieved by replacing some of the clinker with alternatives such as slag, fly ash, and limestone.
- Innovative carbon and novel cement capture technologies: Advances are underway to produce next-generation cement that could be important in realizing a reduction in carbon emissions (Gartner and Hirao 2015). Also referred to as green cement, the products of these technologies are manufactured by using renewable energy and carbon-negative production processes.
- Advanced storage approaches and carbon captures also have the potential in regard to decarbonizing the cement sector (TecEco Pty Ltd. 2013). Current advanced methods and technologies can offer a total reduction of approximately 50% of CO_2 emissions under 2DS by 2050.

Organizations with high management scores have designed strong programs aimed at reducing greenhouse gases with regular monitoring of emissions (Habert 2014). Also, these factories disclose comprehensive metrics on their emissions of scope one and scope two (the emissions from direct operations), with suitable information emissions of scope three (all indirect emissions). In addition, most of the factories with a high management score appreciate the transition threats facing the cement sector (Fig. 3.3) and have accounted for these in their overall risk assessments (Fig. 3.4).

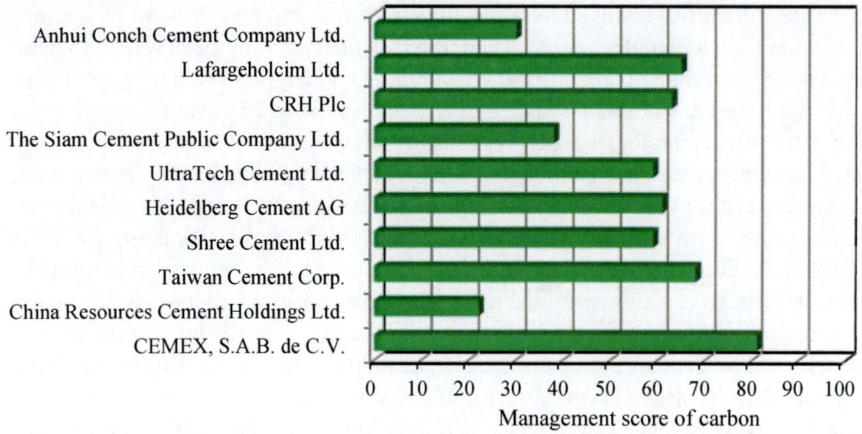

Fig. 3.3 Sustainable analytics data of own operations and management score of carbon (Habert 2014)

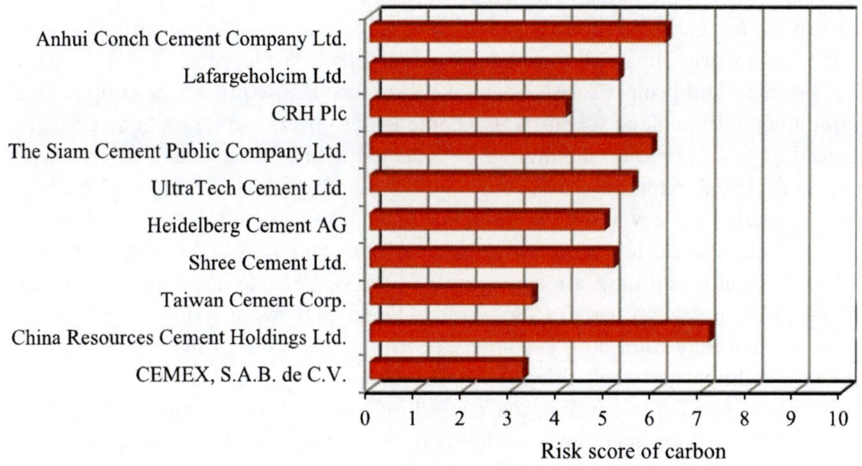

Fig. 3.4 Sustainable analytics data of own operations and risk score of carbon (Habert 2014)

3.4.3 LCC Transition Challenges

Despite increasing interest in LCC technologies and despite the fact that these have advanced considerably, several difficulties remain in relation to embracing them given the difficulties involved in making the transition from one kind of technology to another (Schneider et al. 2011). There is a lack of support of the new technologies from many governments in terms of encouraging cement producers to invest in this area. Also, the cement industry is very conservative in regard to viable implementation of modern technologies. This is because of safety and quality concerns on the part of clients and because it is challenging to negotiate what can be lengthy bureaucratic processes relating to acquiring permits to use alternative fuels in given jurisdictions and to attest to the reliability of new products (Liska and Vandeperre 2008). In addition, building and construction codes and standards vary globally regarding the types of blended cements permitted for construction, posing challenges for engineers utilizing new kinds of concrete and cement. Achieving cost-efficiency which would be achieved by cement producers scaling up and investing in modern technologies remains difficult, as clients do not wish to pay a premium for the new concretes and cements (MPA Cement Fact Sheet 2020). In addition, the lack of legally accepted requirements from regulatory agencies throughout the world in general ensures that the transition process overall is moving slowly.

To achieve sustainability, many countries are focusing on sustainable development by reducing the demand for energy and water and also by recycling multiple waste materials generated by a range of activities, but especially by construction work in the city (World Business Council for Sustainable Development 2009). Sustainable building and project schemes come under sustainable project development in different countries. Such schemes also come under governments' recognized work. According to these projects, buildings constructed with LCC use recycled aluminum and are designed to minimize the consumption of both water and energy. In this way, these countries are developing a green print for sustainable cities in the future and are becoming the world's most sustainable urban communities (Della and Grutzeck 1999). Incubator buildings are an example of the sustainable development projects taking place in several parts of the world. The advantages of living in sustainable cities are that they offer an eco-friendly environment, better growth opportunities, modern facilities, and many things to do that support a high quality of life (Lackner 2003). As construction continues at a rapid rate, this can be a major challenge for people living in cities including in relation to the pollution arising from construction waste. Although there are some pros and cons, the majority of cities are progressing toward achieving sustainable development and becoming eco-friendly, and green cities are developing worldwide.

3.5 Proof of Concepts: Industrial Trials for the Production of LCC

Under strictly controlled conditions, the very encouraging outcomes for ternary blended cement have been achieved on a laboratory scale (Papadakis et al. 1992). Multiple considerations should be taken into account to scale up these processes. Considerable variations in the processes used for the calcination and grinding of the clay between the lab scale and the industrial scale are to be expected (Daspoddar et al. 1999). For manufacturing huge amounts of cement under optimum conditions, a full-scale industrial trial should be conducted. The target ternary cements would have a clinker content of about 50% (Habert et al. 2008). With the co-grinding of the synergetic materials with gypsum and clinker and the homogenizing and mixing of the calcined materials with limestone at a ratio of 2:1, the trials should include the calcination of 100 tons of medium-grade kaolinite clays.

Clay from the Pontezuela clay deposits should be chosen for the trials (Tironi et al. 2013). With an average kaolinite content of 49% measured by thermos-gravimetric analysis (TGA), the authors classify the materials as medium-grade kaolinitic clay. Aided by X-ray fluorescence (XRF), the chemical constituents of the clay samples can be examined as presented in Table 3.2.

Calcination refers to the modification of materials in wet-process rotary kilns, frequently utilized to generate clinker to calcine materials in dry conditions, as illustrated in Fig. 3.5 (Tironi et al. 2013). Clay is fed into the kiln and heated to 750 °C, the optimum temperature for material calcination that 90 tons of calcined material is produced after the calcination (Taylor-Lange et al. 2015). The material loss is

Table 3.2 Chemical compositions of representative Pontezuela clay samples (Tironi et al. 2013)

Oxides	%wt.
Silicon (IV) Oxide	54.60 ± 0.40
AluminumOxide	27.3 ± 1.17
Iron (III) Oxide	12.60 ± 0.07
Calcium Oxide	1.70 ± 0.98
Magnesium Oxide	0.90 ± 0.06
Sulfur (VI) Oxide	0.70 ± 0.07
Sodium Oxide	0.30 ± 0.00
Potassium Oxide	1.60 ± 0.06
Titanium Oxide	0.80 ± 0.00
Phosphorous Oxide	0.20 ± 0.06
Manganese (V) Oxide	0.00 ± 0.00
Cr_2O_3	1.60 ± 0.92
LOI	10.20 ± 0.32
Humidity	4.20 ± 0.07

Fig. 3.5 Direct clay feedings to the rotatory kiln (Taylor-Lange et al. 2015)

related to chemically bound water on the clay structure and the original clay moisture content. The calcined materials are kept in five loads. The calcination quality is examined through the calcined clay dihydroxylation. During calcination, dihydroxylation is achieved when all OH groups are discharged (Fernandez et al. 2011). The individual results of each of the five masses of calcined materials assessed are represented in Fig. 3.6. All the masses have been fully dehydroxlated, which means the materials have been completely stimulated.

The pozzolanic reactivity of the calcined material is examined in relation to the compressive strength of standardized mortar, where pozzolanic materials replace 30 wt% of cement following Antoni and Fernández's protocols. At the laboratories, two series of references are made with 30 wt% pozzolanic material and calcined cement and 100% of ordinary Portland cement (OPC) under optimum conditions (Fernandez et al. 2011). The compressive strength results of standardized mortar at three days, one week, and one month are presented in Fig. 3.7. In the rotary kilns (averages of heaps), the material calcined reactivity proved to be equivalent to that of the materials calcined in laboratory conditions, indicating that industrial calcination is effective.

Under industrial conditions, grinding is carried out by using ball mills with double-chamber grinding systems, as indicated in Fig. 3.8. It is important to grind the materials to avoid low specific surfaces, i.e., to avoid coarseness (Garg and Skibsted 2014). To optimize the alumina phase reaction, the gypsums are modified. The initial gypsums used have a low level of purity. Therefore, the comparatively large proportions in the compositions, but the aggregate content of sulfur (VI) oxide justify the standards of cements. Ultimately, the parameters of the grindings should be set as sulfur (VI) oxide up to 3.0%, an optimum surface of 4000–5000 cm^2/g, and 4–8% retained in the 90 μm sieves (USGS 2012). Grinding should be finished within

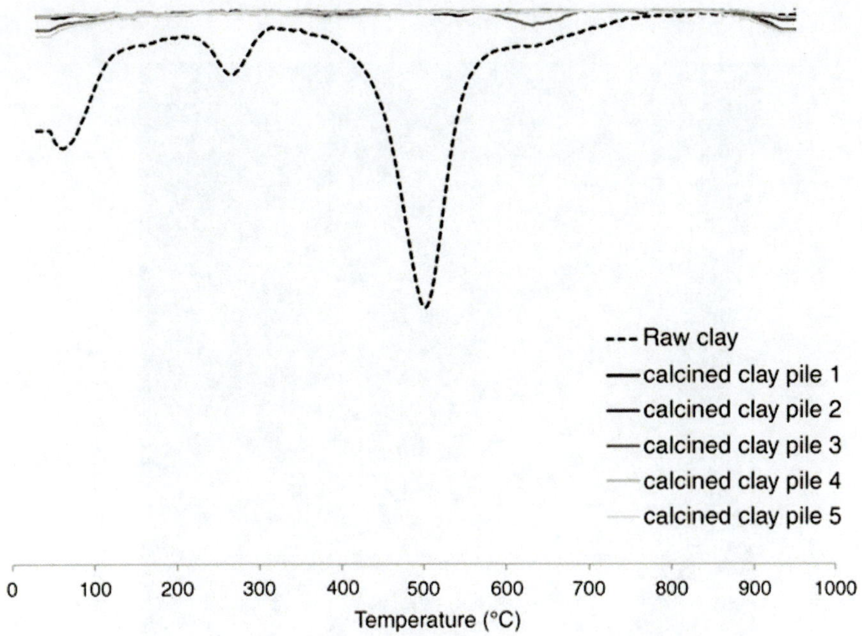

Fig. 3.6 Thermo-gravimetric analysis of the clay calcined in the rotatory kilns (Fernandez et al. 2011)

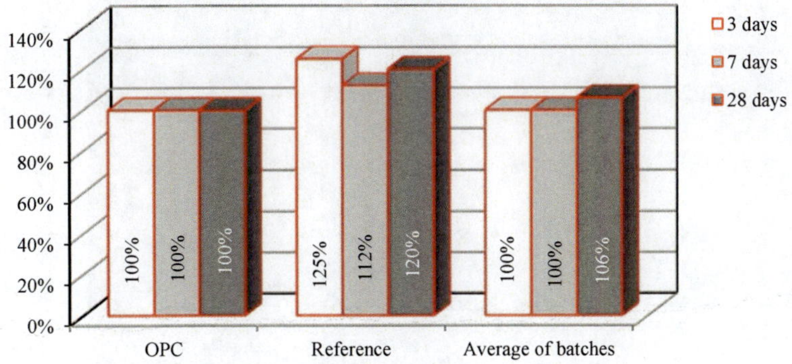

Fig. 3.7 Percent of cement compressive strengths made with 30 wt% of calcined clays standardized to ordinary Portland cements (Fernandez et al. 2011)

eight hours. Material samples should be taken every 30 min (He et al. 2000). The material chemical compositions acquired and the average constituent percentage are presented in Table 3.3.

The final cements are categorized following the protocols developed for blended concrete in global standards and principles. During the process of co-grinding, excess grindings of some softer compounds via interactions with harder materials may

Fig. 3.8 Industrial ball mills utilized to generate the LCCs used to produce through co-grindings (Garg and Skibsted 2014)

Table 3.3 Industrial LCC chemical compositions (He et al. 2000)

Chemical compounds	%wt.
Silicon (IV) oxide	27.3
Aluminum (III) oxide	4.6
Iron (III) oxide	4.6
Calcium oxide	49.8
Magnesium oxide	1.3
Sulfur (VI) oxide	3.7
IR	12.6
LOI	7.1
Calcium oxide$_{free}$	0.9
Total	98.4
Average component Percentages	
Gypsums	8.9
Calcium carbonates/Calcined clays	41.1
Clinkers	50.0

Table 3.4 Mechanical and physical tests' results of the industrial LCCs (Kakali et al. 2001)

Material	Retained No. 4900 Sieve (%)	Consistency (%)	Setting times		Volume Stabilities (mm)	Compressive strength (MPa)		
			Initial (min)	Final (min)		3 days	7 days	30 days
LCCs	12	25	135	174	0.3	11.0	17.5	30.3

be generated (Kakali et al. 2001). This may have an effect on the cement grain size distribution, and it might ultimately increase the mix water demand. Table 3.4 presents the findings and results of the mechanical and physical tests.

3.6 Industrial Uses of Experimental Cement Generated in Industrials Trials

The industrial trials enabled technical teams to ascertain the reaction of the manufacturing and construction sector to the new binders. The cement batches produced were distributed among construction material manufacturers and builders (Habert et al. 2009). A technical team strictly supervised the industrial uses to which the experimental cements were put in the industrial trials. Prospective clients were asked to utilize the new experimental cements in the same percentages as they would expect to use ordinary Portland cement.

Table 3.5 Mix quantities utilized in manufacturing of concretes (Donatello et al. 2010)

For 1 m^3	Gravimetric mix proportions (kg)		Mix proportions (m^3)	
Materials	Hollow blocks 150 mm	Concrete 25 MPa	Hollow blocks 500 × 200 × 150 mm^3	Concrete 5 MPa
LCCs	300	360	1.0	1.0
Quarry of Sand "El Purio"	654	–	1.8	–
Quarry of Powder "Palenques"	–	780	–	1.6
Aggregate 5.0–13.00 mm "El Purio" quarry	1302	–	3.5	–
Aggregates 19.00–10.00 mm "Palenque" quarry	–	1034	–	2.4
Water (L)	112	169	0.4	0.5
Superplasticizers Dynamon SX-32 (L)	–	4	–	–
w/c batches	0.4	0.5	–	–
w/c effective	0.2	0.4	–	–
Designed slumps	0.0	12 ± 2 cm	0	12 ± 2 cm

The trials concentrated on using the cements for two major applications: to produce elements of 25 MPa precast concretes at prefabrication plants and to manufacture blocks of hollow concrete with dimensions of 500 mm^3 × 200 mm^3 × 150 mm^3 manufactured on semi-automated vibro-compacting machines (Donatello et al. 2010). The mix proportions complying with global standards and principles to approve the mix designs are presented in Table 3.5.

Under standard manufacturing conditions, hollow concrete blocks were manufactured, with a ratio of 1:1 cement replacement by the new cements (Scrivener et al. 2018). The quality of the blocks was examined in reference to global standards NC 248:2019. These standards cover water absorption and compressive strength requirements. The results of the analyses of the concrete blocks prepared with LCC and of the reference blocks prepared with ordinary Portland cement are presented in Table 3.6

Table 3.6 Absorptions and compressive strengths of concretes and hollow block made with LCCs (Khatib 2009)

Dimensions of hollow block (mm^3)	Average compressive strengths at 7 days (MPa)	Average compressive strengths at 30 days (MPa)	Performances	Absorption (%)
500 × 200 × 150	3.3	5.9	2.0	5.6
Specification	4.0	5.0	–	≤10.0

Table 3.7 Results of concrete compressive strengths prepared for prefabricated elements with ordinary Portland cement and LCC (Sánchez et al. 2016)

Materials	Cement consumptions (kg/m^3)	Average compressive strength at 3 days (MPa)	Average compressive strengths 7 days (MPa)	Average compressive strengths at 28 days (MPa)	Cement performances[a]
LCC	360	–	21.0	31.4	0.9
Ordinary Portland Cement	360	20.4	–	33.2	0.9

[a]Relationships between the compressive strengths acquired at one month in MPa and consumption of the cements

(Khatib 2009). The water absorption and compressive strength requirements were met by the experimental blocks, thereby showing that the LCCs can be used as a substitute for ordinary Portland cement in this kind of industrial application.

Under standard manufacturing conditions, several cubic meters of 25 MPa concretes were cast with a ratio of 1:1 cement replacement by the new cements. The quality of the precast elements was examined in reference to global standards and principles (Sánchez et al. 2016). The compressive strength results of both the normal and experimental concrete casts in the trials are presented in Table 3.7. No major variations were observed in rheology, and both mixes exceeded the one-month strength recommended, thereby showing that LCC can be used as a substitute for ordinary Portland cement in this kind of industrial application.

3.7 Assessments of the Ecological Impacts of the Experimental Cement Generated in Industrials Trials

The construction industry is one of the fastest-growing industries globally with significant demand for construction supplies (Scrivener et al. 2017), including for concrete in particular. However, as discussed, the production of concrete has a negative impact on the environment. CO_2 emissions and over-exploitation of natural resources are playing a consequential role in this adverse constructional effect on the environment (Harrison 2003). Green concrete prepared with LCC is among the major innovations developed with the goal of curbing the extent of the harm being done to the natural world. However, as recent developments, green concrete and LCC have a long way to go in terms of development and acceptance for practical applications (Scrivener et al. 2017). The characteristics of LCC include great durability along with corrosion-protection properties, appropriateness for fine workmanship, high fire resistance, and strong mechanical properties in relation to strength, shrinkage, creep, and static behaviors.

In the history of the construction industry, LCC and green concrete are advanced technologies and were first presented in Denmark in 1998 under the general concept of "green concrete" (Scrivener 2014). Overall, 8–10% of total CO_2 emissions are the result of cement production. The gas is emitted due to the crushing of limestone and clay and their subsequent burning at extremely high temperatures. There is undoubtedly an urgent need to find a replacement to overcome this problem, and to some extent, the present-day solution is LCC technology (Antoni et al. 2012). In order to be categorized as "green," a concrete must meet two criteria: It must be from a freshly prepared concrete mix, and it must be environmentally friendly. The constituents of this type of concrete are generally reusable/re-manufactured/re-furnished materials, and their use in cement–concrete production generates less CO_2 than does the production of conventional cement concrete. The materials can be recycled demolition waste aggregates, blast furnace slag, fly ash, and recycled glass aggregate (Scrivener et al. 2018). With increased workability and higher ultimate strength compared with conventional cement, LCCs are increasingly being used in the global construction industry.

LCCs are cements made of waste materials to fulfill the purpose of rendering the construction industry more sustainable by reducing its environmental impact (Shi et al. 2011). The waste materials used in this endeavor include industrial waste, fly ash, red mud, silica fume, and blast furnace slag. Agricultural waste such as sugarcane bagasse ash, rice husk ash, and groundnut shell ash is used in the manufacture of sustainable concrete (Maravelaki-kalaitzaki and Moraitou 2000). LCC is used in multiple ways in the construction industry: It is used in road construction and mass construction projects such as for bridges, dams, and retaining walls. Though LCC takes us a step closer to achieving a sustainable construction solution, it does have limitations (Shi et al. 2006), and for this reason, researchers worldwide are working to create a version with properties on a par with those of conventional concrete. When they succeed, it will bring on a revolution in the construction industry.

Cement prepared from raw materials and concrete that is eco-friendly are known as LCCs (Provis 2014). The other way in which an LCC is defined is as a cement-saving structure with minimal ecological impact. The objective of the center for LCC is to minimize the ecological impact of concrete production. New technologies have been developed to achieve this goal. The technologies are developed in reference to all phases of the life cycle of concrete constructions such as specifications, structural design, production, and maintenance. They take into account performance aspects such as ecological aspects (recycling, CO_2, and energy), thermodynamic characteristics (inputs to other features), durability (mechanisms relating to curbing deterioration, frost damage, and corrosion), workmanship (curing, workability, and strength development), fire resistance (heat transfer and spalling), and mechanical characteristics (static, creep, shrinkage, and strength behavior) (Khale and Chaudhary 2007). There are various alternative ecological specifications to which LCC structures must adhere:

- Waste-derived and CO_2 fuels must replace fossil fuels in the production of cement by at least 10%.
- Application of new forms of waste products, formerly disposed of in a landfill or in some other way.
- Application of concrete plants' own waste materials.
- At least 20% of the concrete must be residual materials utilized as aggregates.
- CO_2 emissions must be reduced by at least 30% as compared to the CO_2 emissions associated with the production of conventional concrete.

The production of LCC should fulfill the idea of utilizing eco-friendly materials in cements and concretes to manufacture more sustainable construction materials (De Silva et al. 2007). It describes the use of raw materials in which production processes do not result in ecological damage and are sustainable and result in structures that perform to a high standard over the long term. The three major goals driving the development and production of low-carbon cement and concrete are

- To minimize the cement manufacturing sector's emission of greenhouse gases such as CO_2.
- To minimize the application of natural resources such as natural rocks, natural river sand, shale, clay, and limestone that are currently important in the manufacture of traditional concrete.
- To minimize the application of residual products in concrete that lead to water, air, and land pollution.

LCC is eco-friendly given that its production as compared to that of standard concrete is protective of the environment through the use of industrial residual materials as a partial replacement for cement, natural sand, and coarse aggregates (Somna et al. 2011). It has substantial strength and durability such that it is comparable in terms of these points with conventional concrete. It is exceptionally resistant to corrosion and also withstands acid rains effectively. Buildings constructed with green concrete and LCC are more resistant to temperature changes such that they bring savings in relation to cooling and heating costs (Arjunan et al. 1999). Recyclable and reusable concretes that can be easily transformed and used after their life cycle are used in the process of making green concretes and low-carbon materials. Examples of such materials are recycled concrete aggregate, recycled demolition waste aggregate, glass aggregate, fly ash, blast furnace slag, manufactured sand, rice husk ash, silica fume, and metakaolin (Kovalchuk et al. 2007).

Changes needed in regard to CO_2 emission for the production of LCCs are less compared to ordinary Portland concretes and cements (Rovnaník 2010). In comparison with ordinary Portland cement, LCC has better thermal and acid resistance and compressive and splitting-tensile strength. The production of LCC also consumes less cement than does the production of ordinary Portland cement, thereby reducing CO_2 emissions. Further, LCC is more economical and has better workability than does ordinary Portland concrete. The terms green concrete and low-carbon oxide indicate that a concrete is ecologically sustainable (Jang et al. 2015). Non-biodegradable industrial waste constitutes the principal component of LCC, which means that the production of this kind of cement produces less CO_2 and uses less energy in comparison with the production of ordinary Portland cement. It can be used to completely or partially replace ordinary Portland cement in many applications, thereby reducing overall CO_2 emissions. Most of the industrial wastes used in the production of LCC are in pulverized form, which means that very little energy is used in order to incorporate them into production processes (Majling and Roy 1993). On this basis, challenges related to dumping will be addressed, and water and soil pollution will be dramatically reduced (Barbosa and MacKenzie 2003). If LCC is used to replace ordinary Portland cement to even a modest extent worldwide, energy consumption and global CO_2 emissions will decrease considerably.

With the passage of time, the compressive strength exhibited by LCCs improves considerably (Weil et al. 2009). The low creep and low drying shrinkage of LCCs provide long-term durability. LCCs show great promise for applications in aggressive environments and have excellent resistance to chemical attack, whereas ordinary Portland cement–concrete may not be sufficiently durable in such contexts (Habert et al. 2011). The absence of gypsum in low-calcium fly ash-based LCC and concrete provides a high level of resistance to sulfate attacks. The fire endurance and non-combustible nature of LCC make it superior to any engineering material for construction, transportation, and infrastructure applications. It can withstand temperatures of more than 800 °C (Jang et al. 2014). Thus, LCCs have no major disadvantages in terms of either production or performance as compared to ordinary Portland cement. However, the availability of a labor pool with the needed expertise and the availability of the chemicals needed for production may be concerns.

At present, the green prefix is used for materials and procedures such as green-building, green-bag, and green concrete that have either no effect or a limited effect on the environment and humans relative to conventional production (Park et al. 2018). According to some reports, the cements produced release almost an equivalent amount of carbon dioxide emission of cement (1 ton of cement produces 1 ton of CO_2). With new experiments, various alternatives are being developed that are environmentally friendly and do not entail a compromise in regard to the performance of the structures in which they are used (Xie et al. 2003). Cements are replaced by fly ash up to 30% due to the pozzolana property of the latter such that it behaves similarly to cement as a binding material. Further, LCCs have low heat of hydration and produce only a small amount of CO_2. Today, most of the cements on the market are fly ash-based. However, the new cement technique could be entirely replaced with LCC (fly ash + alkali medium + sodium silicate or more different composite) (Barcelo et al. 2014). Low carbon-based cements are 30% more environmentally friendly than ordinary Portland cement (Ferna 2002). They have low heat of hydration, low creep and shrinkage, and good fire and thermal resistance.

With clinker replacement of up to 30%, blended cements enable a reduction of approximately 14–22% of the emissions of CO_2 associated with the production of ordinary Portland cement (Gartner and Sui 2018). In LCC formulations, a larger proportion of clinker than in ordinary Portland cement can be used—that is, 50% of the clinker can be replaced with alternative materials without affecting performance (Gartner 2017). These figures represent a reduction of approximately 30% of the

Table 3.8 Emissions of CO_2 versus clinker factors in the manufacturing of cements (Sui et al. 2006)

Phases of the productive processes	Unitaryvalues (kg oxidecalcium /ton)	Clinker factors (%)		
		100% Clinkers	70% Clinkers	55% Clinkers
Calcination ofRaw materials (Calcium and MagnesiumOxide Oxide)	502	502	351.4	276.1
Fuels	320	320	224	176
Addition (calcined clays and limestones)	380	0	38.2	57.2
Grindings	100	100	100	100
Others	60	60	60	60
Total		982	773.6	669.3
Savings associated with clinker factors of 100%			79%	68%

emissions of CO_2 related to the manufacturing of cement, as shown in Table 3.8 (Sui et al. 2006). Based on the industrial trials at cement factories and production viability from an environmental viewpoint, preliminary assessments of the environmental impacts were made.

Clinker production facilities rely on wet-process rotary kilns (Sui et al. 2015). However, factories are no longer using clinker produced in this way as it is not very efficient. The clinkers utilized now are produced in high-efficiency dry rotary kilns. Also, these latter kilns allow clay calcination at cement factories in retrofitted cement kilns for wet processes (Diao 2008). This might allow present production infrastructure such as clay calcination, cement grinding, and storage in LCC industries and yet still improve efficiency and enable gains in terms of protecting the environment (Sui and Liu 1999). Calcination of clays can take place in two ways:

- Short-term: Calcined clay production occurs via existing rotary kilns with a wet process in LCC factories.
- Long-term: Calcined clay production occurs in large-scale retrofitted calciners with a dry process in LCC factories.

During the production of cement, CO_2 emissions are computed by dividing the productive processes into two stages: (i) production of clinker and calcination of clay and (ii) grinding and mixing of additive materials and clinker (Sui and Liu 1999). Three major sources of emissions can be described during the calcination process: (a) the combustion of fossil fuels; (b) the chemical decomposition of calcium carbonate and magnesium carbonate, and (c) the consumption of energy. The methods to approximate the emissions of CO_2 are in compliance with the 2019

Table 3.9 Emissions of CO_2 because of the consumption of energy during generation of clinkers cement production factories (Taylor 1990)

Consumption indices (energy units/time)	Emission factors (kg CO_2/energy unit) [a]	Emissions of CO_2 (kg CO_2/ton)
Emissions because of consumptions of fuels (pet-cokes)		
100 (kg pet-cokes)	4.08 (kilogram CO_2/kg pet-cokes)	409.00
Emission because of consumptions of electricityGES		
0.03890	745 (kg CO_2/MWh)	30.78
Aggregate CO_2 emitted per consumption of energy one ton of clinker		437.67

Climatic Change Intergovernmental Panel recommendations, whereas the method to measure the emissions of CO_2 arising from the production of clinker is multiplied by the factor of emissions. CO_2 emissions caused by the consumption of energy (a and c) are computed applying the methodology expressed as Eq. (3.10) (Zhang et al. 1999):

$$CO_2 \text{ emissions} = \text{Emission factor} \times \text{Consumption index} \tag{3.10}$$

In which the emissions of carbon oxide are as follows: Emission factor of kilogram CO_2/ton clinker-calcined clays: Consumption indices of kilogram CO_2 per energy unit consumed: energy unit utilized per clinker-calcined clay tons. The global energy system (GES) supplies the clinker used in the dry process with the electricity and pet-coke as the source of energy (Taylor 1990). Figures for the emissions related to the consumption of energy in the first cement production phase during which a large amount of clinker is produced are presented in Table 3.9.

Electricity was supplied by the global energy system, and crude oil was used for the calcination of clay at the cement manufacturing plants (Chatterjee 2002). The emissions related to the consumption of energy during the calcination of clay over short-term and long-term periods are shown in Table 3.10.

The chemical decomposition of the calcium carbonate and magnesium carbonate included in the composition of the raw materials (clay and gray limestone) cause the emission of CO_2 (Older 2000). The emissions can be inferred through stoichiometry computations using two different methods:

- By computing variations in the calcium oxide and magnesium oxide content of the raw materials before being fed into the kilns and the exits (clinker or calcined clay).
- By examining the contributions of all paste components to the aggregate emissions of CO_2 via measurements of the chemical composition of the minerals utilized in the concrete at the entrances of the kilns and their magnesium carbonate and calcium carbonate content.

Table 3.10 Emissions of CO_2 because of the consumption of energy during calcinations of clays in LCC factories (Chatterjee 2002)

Scenarios	Consumption indices(energy unit/time)	Emission factors(kg CO_2/energy unit)	Emission of CO_2(kg CO_2/time)
Emission because of combustions of fuels especially crude oils			
Short-term (rotatory kiln-wet processes)	105.5[a] (kg crude oil)	3.20 (kg CO_2/kg crude oil)	342.66
Long-term (industrial calciner)	58.2[b]	3.20 (kg CO_2/kg crude oil)	187.72
Emission because of GES electricity consumptions			
Short-term (rotatory kiln-wet processes)	0.0399[a]	744 (kg CO_2/MWh)	29.79
Long-term (industrial calciners)	0.0238[b]	744 (kg CO_2/MWh)	17.71

[a]Reductionsof consumption of global crude oils for the generations of calcined clays associated with production of clinkers are approximated at forty percent, because of the reduction of the calcination temperatures and a decrease of 16% of consumption of electricity since clays do not need to be processed in the paste mills (Valderrama et al. 2012)
[b]Reductions of 45% of the consumptions of global crude oils and 40% of the consumption of electricity are approximated associated with the productions of wet processes in the rotary kilns, because of the residence time in the calciners and calcination temperature reductions (the materials are introduced with natural humidity) (Vizcaíno-Andrés et al. 2015)

The available data determine which approach is applied to explaining the chemical composition of the materials before and after processing (Lan and Glasser 1996). Researchers measure the magnesium oxide and calcium oxide content of the materials at the point of entering the kilns and exiting the kilns to establish differences between the content. The stoichiometry of the CO_2 emitted provides a way of measuring variations in the content of the oxides (Mehta 1973). The CO_2 emissions because of the limestone decarbonation were computed for LCC factories that are major suppliers of clinker for cement production (Older 2000). The calculations used to measure CO_2 emissions at the point of magnesium and calcium carbonate decarbonation during the generation of clinker at cement plants are presented in Table 3.11.

Table 3.11 Data and formulas utilized for the computations of CO_2 emitted because of the clinker production by dry processes (Older 2000)

CO_2 emitted (calcinations) = [0.7850 × (Out calcium oxide-in calcium oxide) + 1.0920 × (out magnesium oxide-in magnesium oxide)]/100

Out calcium oxide (%)	In calcium oxide (%)	Out magnesium oxide (%)	In magnesium oxide (%)	CO_2 emitted for de-carbonizations (ton calcium oxide/ton clinker)
65.00	0.00	1.63	0.00	0.54

Table 3.12 Emission of CO_2 during grinding of cements (Su and Kurdowski 1992)

Production	Consumption indices MWh/ton)	Emission factors (kilogram CO_2/MWh)	Emissions of CO_2 (kilogram CO_2/t)
Clinkers/calcined Clays[a]	0.0466	745	35.6

[a]It was considered that the mixing and grindings of the various cement types have the same consumption of energy

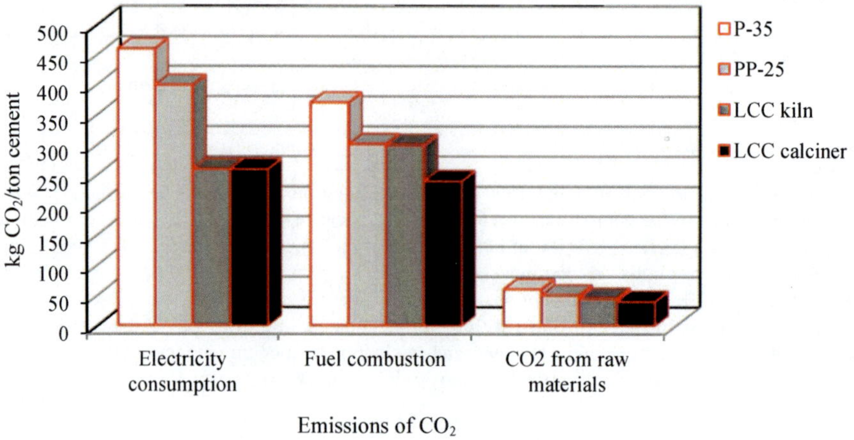

Fig. 3.9 Emissions of CO_2 related with production of LCCs with 45% of supplementary concrete materials and in comparison with three reference cements (Beretka et al. 1997)

CO_2 emissions in the cement grinding phases are related to the consumption of electricity in the ball mills as shown in Table 3.12 (Su and Kurdowski 1992). Other emission sources are discarded because they are considered negligible.

Figure 3.9 presents the final calculations of emissions of CO_2. During the industrial trials, the LCCs manufactured in non-optimal conditions showed a decrease in CO_2 to about 270 kg CO_2/ton as compared with ordinary Portland cement (P-35). This is a decrease of about 30% (Beretka et al. 1997). In reference to conventional global blended cement (PP-25), the decrease in CO_2 emissions as compared to the CO_2 emissions associated with ordinary Portland cement is in the range of 125 kg CO_2/ton. Both PP-25 and P-35 are standard products generated at LCC plants.

The approximate aggregate emissions for LCCs achieved an emissions reduction of more than 200 kg CO_2/ton for the production of PP-25 and of more than 300 kg CO_2/ton for P-35 (Factsheet 2014). In comparison with business as usual practices, these figures represent an overall reduction of more than 35% over emissions of CO_2 arising from the production of standard concrete. The compressive strength and emissions at one month of LCCs generated at the industrial trials are presented in Fig. 3.10, in comparison with the cement reference values of PP-25 and P-35 (Ambroise 2008). These data clearly show that even in comparison with other

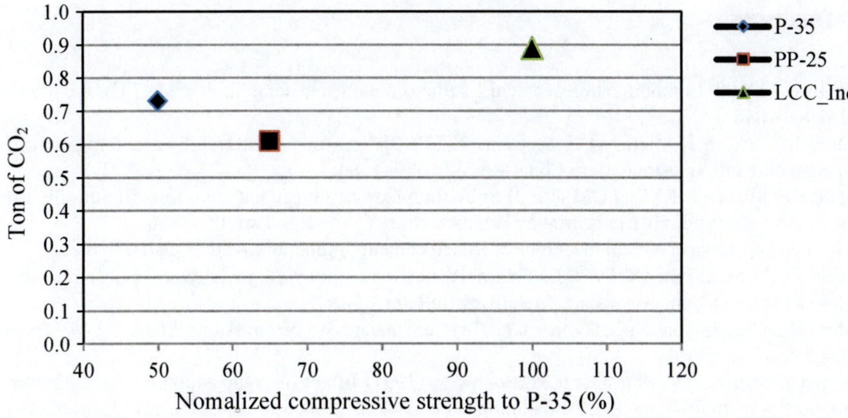

Fig. 3.10 Relationships between CO_2 emission/ton cement and related compressive strengths of LCC, P-35 and PP-25 generated in the industrial trials (Ambroise 2008)

cements industrially generated at cement factories LCCs do not compromise cement performance despite their very low clinker content (from which their low emission of CO_2 results).

3.8 Conclusions

Through the application of ternary systems based on clinker, limestone, and calcined clay, there are alternatives available to replace clinker in the production of blended cement. Underlying this review, the defining principle of the advanced technologies is that the synergy between limestone and calcined clay makes it possible to decrease the clinker used, thereby improving the reactivity of the supplementary concrete material. These systems are founded on the application of medium-grade kaolinite clay. Only small changes are needed in the processes of production in order to make use of the replacement materials. Reserves of medium-grade limestone and clay are much greater than those of any other supplementary concrete materials. Industrial trials to generate cement with clinker factors of 51% at an industrial scale have confirmed that the new systems are very strong, and even that acceptable results have accrued in non-optimal conditions in regard to the resulting performance for these materials used in concrete. The new concrete systems could enable a reduction in CO_2 emissions in the range of 24–35% related to the production of cement. This reduction would be based on substituting clinker, which is the major emitter of CO_2, with combinations of substances that emit very little CO_2 in processing, production, or use.

References

Ambroise J (2008) Immobilization of calcium sulfate contained in demolition waste. J Hazard Mater 151:840–846

Antoni M, Rossen J, Martirena F, Scrivener K (2012) Cement substitution by a combination of metakaolin and limestone. Cem Concr Res 42:1579–1589

Arjunan P, Silsbee MR, Roy DM (1999) Sulfoaluminate-belite cement from low-calcium fly ash and sulfur-rich and other industrial by-products. Cem Concr Res 29:1305–1311

ATILH (2002) Environmental inventory of french cement production. ATILH, Paris, France

Barbosa VFF, MacKenzie KJD (2003) Thermal behaviour of inorganic geopolymers and composites derived from sodium polysialate. Mater Res Bull 38:319–331

Barcelo L, Kline J, Walenta G, Gartner E (2014) Cement and carbon emissions. Mater Sci 47:1055–1065

Beretka J, Sherman N, Marrocooli M, Pompo A (1997) Effect of composition on the hydration properties of rapid-hardening silfoaluminate cements. In: Proceedings of the 10th international congress on the chemistry of cement, Gothenburg, Sweden, 2–6 June 1997

Boesch ME, Hellweg S (2010) Identifying improvement potentials in cement production with life cycle assessment. Environ Sci Technol 44:9143–9149

Bontempi E (2017) A new approach for evaluating the sustainability of raw materials substitution based on embodied energy and the CO_2 footprint. J Clean Prod 162:162–169

Bontempi E (2017) A new approach to evaluate the sustainability of raw materials substitution. In: Raw materials substitution sustainability. Springer, Cham, Switzerland. ISBN 978-3-319-60830-3

Bosoaga A, Masek O, Oakey JE (2009) CO_2 capture technologies for cement industry. Energ Procedia 1:133–140

Cao Z, Shen L, Zhao J, Liu, L, Zhong S, Yang Y (2016) Modeling the dynamic mechanism between cement CO_2 emissions and clinker quality to realize low-carbon cement. Resour Conserv Recycl 113:116–126

Chatterjee AK (2002) Structure and performance cements. In: Bensted J, Barnes P (eds) Special cements. E & FN Spon, London, UK, pp 226–231

Chen C, Habert G, Bouzidi Y, Jullien A (2010) Environmental impact of cement production: detail of the different processes and cement plant variability evaluation. J Clean Prod 18:478–485

Crossin E (2015) The greenhouse gas implications of using ground granulated blast furnace slag as a cement substitute. J Clean Prod 95:101–108

Damtoft JS, Herfort D, Yde E (2006) Concrete binders, mineral additions and chemical admixtures: state of the art and challenges for the 21st century. University of Dundee, Scotland, UK

Damtoft JS, Lukasik J, Herfort D, Sorrentino D, Gartner, EM (2008) Sustainable development and climate change initiatives Original Research Article. Cem Concr Res 38:115–127

Daspoddar K, Chowdhry BN, Basu R (1999) Effect of calcination temperature of magnesium carbonate on the properties of magnesium oxysulphate cement. J Indian Chem Soc 550–551

De Silva P, Sagoe-Crenstil K, Sirivivatnanon V (2007) Kinetics of geopolymerization: role of Al_2O_3 and SiO_2. Cem Concr Res 37:512–518

De Weerdt K, Ben Haha M, Le Saout G, Kjellsen KO, Justnes H, Lothenbach B (2001) Hydration mechanisms of ternary Portland cements containing limestone powder and fly ash. Cem Concr Res 41:279–291

Deja J, Uliasz-Bochenczyk A, Mokrzycki E (2010) CO_2 emissions from the Polish cement industry. Int J Greenhouse Gas Control 4:583–588

Della ES, Grutzeck MW (1999) Giga-scale disposal: a real frontier for ceramic research. Mater Res Innov 3:55–56

Diao J (2008) Actuality and perspective of Chinese CSA development (in Chinese). J Chin Concr Cem Assoc

Donatello S, Tyrer M, Cheeseman CR (2010) Comparison of test methods to assess pozzolanic activity. Cem Concr Compos 32:121–127

European Commission DG (2018) A new approach to evaluating the sustainability of substituting raw materials: science for environmental policy. The University of the West of England, Bristol, UK

Factsheet MP (2014) Mineral planning factsheet. Available online https://www.bgs.ac.uk/downloads/start.cfm?id=1353. Accessed on 1 October 2020

Febelcem VZW (2006) Environmental report of the Belgian cement industry. Febelcem v z w, Brussels, Belgium

Ferna A (2002) The alkali–Silica reaction in alkali-activated granulated slag mortars with reactive aggregate. Cem Concr Res 32:1019–1024

Fernandez R, Martirena F, Scrivener KL (2011) The origin of the pozzolanic activity of calcined clay minerals: a comparison between kaolinite, illite and montmorillonite. Cem Concr Res 41:113–122

Flower DJM, Sanjayan JG (2007) Greenhouse gas emissions due to concrete manufacture. Int J Life Cycle Assess 12:282

Fodor Z, Klemeš JJ (2012) Waste as an alternative fuel—minimising emissions and effluents by advanced design. Process Saf Environ Prot 90:263–284

Gäbel K, Forsberg P. Tillman AM, (2004) The design and building of a lifecycle-based process model for simulating environmental performance, product performance and cost in cement manufacturing. J Clean Prod 12:77–93

Galvez-Martos JL, Schoenberger H (2014) An analysis of the use of life cycle assessment for waste co-incineration in cement kilns. Resour Conserv Recycl 86:118–131

Garg N, Skibsted J (2014) Thermal activation of a pure montmorillonite clay and its reactivity in cementitious systems. J Phys Chem 118:11464–11477

Gartner E (2004) Industrially interesting approaches to "low-CO_2" cements. Cem Concr Res 34:1489–1498

Gartner E (2017) What are BYF cements, and how do they differ from CSA cements? In: Proceedings of the future of cement, 200 years after Louis Vicat. UNESCO, Paris, France, 6–8 June 2017

Gartner E, Hirao H (2015) A review of alternative approaches to the reduction of CO_2 emissions associated with the manufacture of the binder phase in concrete. Cem Concr Res 78:126–142

Gartner E, Sui T (2018) Alternative cement clinkers. Cem Concr Res 114:27–39

Gartner EM, MacPhee DE (2011) A physico-chemical basis for novel cementitious binders. Cem Concr Res 41:736–749

Georgiopoulou M, Lyberatos G (2018) Life cycle assessment of the use of alternative fuels in cement kilns: a case study. J Environ Manag 216:224–234

Habert G (2014) Assessing the environmental impact of conventional and 'green' cement production. Eco-efficient Constr Build Mater 199–238

Habert G, Billard C, Rossi P, Chen C, Roussel N (2010) Cement production technology improvement compared to factor 4 objectives. Cem Concr Res 40:820–826

Habert G, Choupay N, Escadeillas G, Guillaume D, Montel JM (2009) Clay content of argillites: influence on cement based mortars. Appl Clay Sci 43:322–330

Habert G, Choupay N, Montel JM, Guillaume D, Escadeillas G (2008) Effects of the secondary minerals of the natural pozzolans on their pozzolanic activity. Cem Concr Res 38:963–975

Habert G, Espinose JB, Lacaillerie D, Roussel N (2011) An environmental evaluation of geopolymer based concrete production: reviewing current research trends. J Clean Prod 19:1229–1238

Harrison AJW (2003) New cement based on the addition of reactive magnesia to Portland cement with or without added pozzolana. In: Proceedings of the CIA conference: concrete in the Third Millennium, Brisbane, Australia, 17–19 July 2003

Hasanbeigi A, Price L, Lu H, Lan W (2010) Analysis of energy-efficiency opportunities for the cement industry in Shandong Province, China: a case study of 16 cement plants. Energy 35:3461–3473

He C, Makovicky E, Osbæck B (2000) Thermal stability and pozzolanic activity of raw and calcined mixed-layer mica/smectite. Appl Clay Sci 17:141–161

Hendriks CA, Worrell E, de Jager D, Block K, Riemer P (2004) Emission reduction of greenhouse gases from the cement industry. In: Proceedings of the IEA greenhouse gas control technologies conference, Vancouver, BC, Canada, 5–9 September 2004

Huntzinger DN, Eatmon TD (2009) A life-cycle assessment of Portland cement manufacturing: comparing the traditional process with alternative technologies. J Clean Prod 17:668–675

Imbabi MS, Carrigan C, McKenna S (2012) Trends and developments in green cement and concrete technology. Int J Sustain Built Environ 1:194–216

ISO (2006) 14040: environmental management-life cycle assessment-principles and framwork. ISO, London, UK

ISO (2006) 14044: environmental management-life cycle assessment-requirements and guidelines. ISO, London, UK

Italcementi Group (2012) Environmental product declaration as per ISO 14025 and EN 15804. Institut Bauen und Umwelt e.V. (IBU), Berlin, Germany. Available online https://epd-online.com/EmbeddedEpdList/Download/9128. Accessed on 2 October 2020

Jang JG, Ahn YB, Souri H, Lee HK (2015) A novel eco-friendly porous concrete fabricated with coal ash and geopolymeric binder: heavy metal leaching characteristics and compressive strength. Constr Build Mater 79:173–181

Jang JG, Lee NK, Lee HK (2014) Fresh and hardened properties of alkali-activated fly ash/slag pastes with superplasticizers. Constr Build Mater 50:169–176

Josa A, Aguado A, Cardim A, Byars E (2007) Comparative analysis of the life cycle impact assessment of available cement inventories in the EU. Cem Concr Res 37:781–788

Josa A, Aguado A, Heino A, Byars E, Cardim A (2004) Comparative analysis of available life cycle inventories of cement in the EU. Cem Concr Res 34:1313–1320

Kakali G, Perraki T, Tsivilis S, Badogiannis E (2001) Thermal treatment of kaolin: the effect of mineralogy on the pozzolanic activity. Appl Clay Sci 20:73–80

Kartini K, Hamidah MS, Norhana AR, Nur Hanani AR (2014) Quarry dust fine powder as substitute for ordinary Portland cement in concrete mix. J Eng Sci Technol 9:191–205

Khale D, Chaudhary R (2007) Mechanism of geopolymerization and factors influencing its development: a review. J Mater Sci 42:729–746

Khatib JM (2009) Sustainability of construction materials. Woodhead Publishing Limited, Cambridge, UK

Kovalchuk G, Fernández-Jiménez A, Palomo A (2007) Alkali-activated fly ash: effect of thermal curing conditions on mechanical and microstructural development—Part II. Fuel 86:315–322

Lackner KS (2003) Erratum: a guide to CO_2 sequestration. Science 300:1677–1678. Available online http://science.sciencemag.org/content/300/5626/1677.long. Accessed on 2 October 2020)

Lan W, Glasser FP (1996) Hydration of calcium sulphoaluminate cements. Adv Cem Res 8:127–134

Li C, Sun H, Li L (2010) A review: the comparison between alkali-activated slag (Si + Ca) and metakaolin (Si + Al) cements. Cem Concr Res 40:1341–1349

Liska M, Vandeperre LJ (2008) Influence of carbonation on the properties of reactive magnesia cement-based pressed masonry units. Adv Cem Res 20:53–64

Ludwig HM, Zhang W (2015) Research review of cement clinker chemistry. Cem Concr Res 78:24–37

Lund P (2007) Impacts of EU carbon emission trade directive on energy-intensive industries—indicative micro-economic analyses. Ecol Econ 63:799–806

Majling J, Roy DM (1993) The potential of fly ash for cement manufacture. Am Ceram Soc Bull 72:45–50

Maravelaki-kalaitzaki P, Moraitou G (2000) Sorel's cement mortars Decay susceptibility and effect on Pentelic marble. Cem Concr Res 29:1929–1935

Mehta PK (1973) Mechanism of expansion associated with ettringite formation. Cem Concr Res

Mehta PK, Monteiro PJM (2006) Concrete microstructure, properties, and materials, 3rd ed. McGraw-Hill, New York, NY, USA. ISBN 0071589198

Mikulčić H, Vujanović M, Duić N (2013) Reducing the CO_2 emissions in Croatian cement industry. Appl Energ 101:41–48

Mindess S, Young JF, Darwin D (2003) Concrete. Prentice Hall, Pearson Education, Inc, Upper Saddle River, NJ, USA, p 644. ISBN 9780130646323

Moesgaard M, Herfort D, Steenberg M, Kirkegaard LF, Yue Y (2011) Physical performances of blended cements containing calcium aluminosilicate glass powder and limestone. Cem Concr Res 41:359–364

Moya JA, Pardo N, Mercier A (2011) The potential for improvements in energy efficiency and CO_2 emissions in the EU27 cement industry and the relationship with the capital budgeting decision criteria. J Clean Prod 19:1207–1215

MPA Cement Fact Sheet (2020) Novel Cements: low energy, LCCs. pp 1–9. Available online https://cement.mineralproducts.org/documents/. Accessed on 6 October 2020

Ojan M, Montenegro P, Borsa M, Altert C, Fielding R (2016) Development of new types of LCC. Available online https://www.wbcsd.org/Sector-Projects/Cement-Sustainability-Initiative/News/CSI-climate-and-energy-workshop. Accessed on 4 October 2020

Older I (2000) Cements containing calcium sulfoaluminate. In: Bentur A, Mindess S (eds) Special inorganic cements. E & FN Spon, London, UK. pp 69–87

Pacheco-Torgal F, Castro-Gomes J, Jalali S (2008) Alkali-activated binders: a review. Part 1. Historical background, terminology, reaction mechanisms and hydration products. Constr Build Mater 22:1305–1314

Papadakis VG, Fardis MN, Vayenas CG (1992) Hydration and carbonation of pozzolanic cements. Mater J 89:109–130

Park SM, Jang JG, Lee HK (2018) Unlocking the role of MgO in the carbonation of alkali-activated slag cement. Inorg Chem Front 5:1661–1670

Pennington DW, Potting J, Finnveden G, Lindeijer E, Jolliet O, Rydberg T, Rebitzer G (2004) Life cycle assessment Part 2: current impact assessment practice. Environ Int 30:721–739

Phair JW (2006) Green chemistry for sustainable cement production and use. Green Chem 8:763–780

Provis JL (2014) Geopolymers and other alkali activated materials: why, how, and what? Mater Struct 47:11–25

Rashad AM (2015) An exploratory study on high-volume fly ash concrete incorporating silica fume subjected to thermal loads. J Clean Prod 87:735–744

Rashad AM, Bai Y, Basheer PAM, Milestone NB, Collier NC (2013) Hydration and properties of sodium sulfate activated slag. Cem Concr Compos 37:20–29

Rashad AM, Zeedan SR (2011) The effect of activator concentration on the residual strength of alkali-activated fly ash pastes subjected to thermal load. Constr Build Mater 25:3098–3107

Rebitzer G, Ekvall T, Frischknecht R, Hunkeler D, Norris G, Rydberg T, Schmidt WP, Suh S, Weidema BP, Pennington DW (2004) Life cycle assessment Part 1: framework, goal and scope definition, inventory analysis, and applications. Environ Int 30:701–720

Rovira J, Mari M, Nadal M, Schuhmacher M, Domingo JL (2010) Science of the total environment partial replacement of fossil fuel in a cement plant: risk assessment for the population living in the neighborhood. Sci Total Environ 408:5372–5380

Rovnaník P (2010) Effect of curing temperature on the development of hard structure of metakaolin-based geopolymer. Constr Build Mater 24:1176–1183

Sánchez Berriel S, Favier A, Rosa Domínguez E, Sánchez MacHado, IR, Heierli U, Scrivener K, Martirena Hernández F, Habert G (2016) Assessing the environmental and economic potential of Limestone Calcined Clay Cement in Cuba. J Clean Prod 124:361–369

Schneider M, Romer M, Tschudin M, Bolio H (2011) Sustainable cement production-present and future. Cem Concr Res 41:642–650

Scrivener K, Martirena F, Bishnoi S, Maity S (2017) Calcined clay limestone cements (LC^3). Cem Concr Res 114:49–56

Scrivener KL (2014) 202 special issue—options for the future of cement. Indian Concr J 88:11–21

Scrivener KL (2016) Eco-efficient cements: potential economically viable solutions for a low-CO_2 cement-based materials industry. Cem Concr Res 114:2–26

Scrivener KL, Avet F, Maraghechi H, Zunino F, Ston J, Favier A, Hanpongpun W (2018) Impacting factors and properties of limestone calcined clay cements (LC^3). In: Green materials. ICE Publishing, London

Scrivener KL, John VM Gartner EM (2018) Eco-efficient cements: potential economically viable solutions for a low-CO_2 cement-based materials industry. Cem Concr Res

Shi C, Jiménez AF, Palomo A (2011) New cements for the 21st century: the pursuit of an alternative to Portland cement. Cem Concr Res 41:750–763

Shi C, Krivenko PV, Roy D (2006) Alkali-activated cements and concrete. Taylor & Francis, New York, NY, USA. ISBN 978-0-415-70004-7

Smith RA, Kersey JR, Griffiths PJ (2002) The construction industry mass balance: resource use, wastes and emissions. Construction 4:680

Somna K, Jaturapitakkul C, Kajitvichyanukul P, Chindaprasirt, P (2011) NaOH-activated ground fly ash geopolymer cured at ambient temperature. Fuel 90:2118–2124

Stafford FN, Raupp-Pereira F, Labrincha JA, Hotza D (2016) Life cycle assessment of the production of cement: a Brazilian case study. J Clean Prod 137:1293–1299

Stemmermann P, Beuchle G, Garbev K, Schweike UC (2011) A new sustainable hydraulic binder based on calcium hydrosilicates. In: Proceedings of the 13th international congress on the chemistry of cement, Madrid Spain, 13 June 2011

Stemmermann P, Schweike U, Garbev K, Beuchle G (2010) Celitement—a sustainable prospect for the cement industry. Cem Int 8:52–66

Su M, Kurdowski W (1992) Development in non-Portland cements. In: Proceedings of the 9th international congress on the chemistry of cement, New Dehli, India. 317–354. Available online https://catalog.hathitrust.org/Record/009223995. Accessed on 8 October 2020

Sui T, Fan L, Wen Z, Wang J (2015) Properties of belite-rich Portland cement and concrete in China. J Civil Eng Archit 9:384–392

Sui T, Li J, Peng X, Li W, Wen Z, Wang J, Fan L (2006) A comparison of HBC & MHC massive concretes for three gorges project in China. In: Measuring, monitoring and modeling concrete properties. Springer, Dordrecht, The Netherlands, pp 341–342

Sui T, Liu K (1999) Study on the properties of high-belite cement. J Chin Chem Soc 488–492

Taylor HFW (1990) Cement chemistry. Academic Press, London, UK

Taylor-Lange SC, Lamon EL, Riding KA, Juenger MCG (2015) Calcined kaolinite-bentonite clay blends as supplementary cementitious materials. Appl Clay Sci 108:84–93

TecEco Pty Ltd (2013) TecEco Cements. TecEco Pty Ltd, Glenorchy, Australia

Tironi A, Trezza MA, Scian AN, Irassar EF (2013) Assessment of pozzolanic activity of different calcined clays. Cem Concr Compos 37:319–327

Turner LK, Collins FG (2013) CO_2 equivalent (CO_{2-e}) emissions: a comparison between geopolymer and OPC cement concrete. Constr Build Mater 43:125–130

US (2018) Geological survey. Mineral commodity summaries. U.S. Geological Survey, Reston, VA, USA

USGS United States Geological Survey (2012) Available online https://minerals.usgs.gov/minerals. Accessed on 2 October 2020

Valderrama C, Granados R, Cortina JL, Gasol CM, Guillem M, Josa A (2012) Implementation of best available techniques in cement manufacturing: a life-cycle assessment study. J Clean Prod 25:60–67

Valipour M, Yekkalar M, Shekarchi M, Panahi S (2014) Environmental assessment of green concrete containing natural zeolite on the global warming index in marine environments. J Clean Prod 65:418–423

Van Den Heede P, De Belie N (2012) Environmental impact and life cycle assessment (LCA) of traditional and "green" concretes: literature review and theoretical calculations. Cem Concr Compos 34:431–442

Van Oss HG, Padovani AC (2003) Cement manufacture and the environment, Part II: environmental challenges and opportunities. J Ind Ecol 7:93–126

Vizcaíno-Andrés LM, Sánchez-Berriel S, Damas-Carrera S, Pérez-Hernández A, Scrivener KL, Martirena-Hernández JF (2015) Industrial trial to produce a low clinker, low carbon cement. Materiales de Construcción 65(317):04

Von Bahr B, Hanssen OJ, Vold M, Pott G, Stoltenberg-Hansson E, Steen B (2003) Experiences of environmental performance evaluation in the cement industry. Data quality of environmental performance indicators as a limiting factor for benchmarking and rating. J Clean Prod 11:713–725

Weil M, Dombrowski K, Buchwald A (2009) Life-cycle analysis of geopolymers. In: Geopolymers. Woodhead Publishing Limited, Cambridge, UK, pp 194–210

World Business Council for Sustainable Development (2002) Toward a sustainable cement industry, March 2002. Available online http://citeseerx.ist.psu.edu/viewdoc/. Accessed on 12 October 2020

World Business Council for Sustainable Development (2009) Cement technology roadmap 2009, carbon emissions reductions up to 2050. Available online https://www.iea.org/publications/freepublications/publication/Cement.pdf. Accessed on 6 November 2018

World Business Council for Sustainable Development (2009) The cement sustainability initiative, cement industry energy and CO_2 performance: "Getting the Numbers Right", vol 44. WBCSD, Geneva, Switzerland. ISBN 978-3-940388-48-3

Xie Z, Xiang W, Xi Y (2003) ASR potentials of glass aggregates in water-glass activated fly ash and Portland cement mortars. J Mater Civil Eng 15:67–74

Yang KH, Jung YB, Cho MS, Tae SH (2016) Effect of supplementary cementitious materials on reduction of CO_2 emissions from concrete. Handb Low Carbon Concr 103:89–110

Zhang L, Muzhen S, Yanmou W (1999) Development and use of sulfo and ferro aluminate cements in China. Adv Cem Res 11:15–21

Chapter 4
Principles of Fiber-Reinforced Concrete

Abstract It is possible to enhance the ability of crack control and inherent brittleness of ordinary concretes by integrating discrete fibers into concretes. Fiber-reinforced concretes are acknowledged as high-performance building materials due to their high levels of toughness under tensile and compressive loads. It is therefore broadly applied in precast structures, bridges, tunnels, and high-rise buildings. Societal demand has raised the requirements for advanced fiber-reinforced concrete composites with multifunctionality and ultra-high performance such as self-regulating, self-clearing, self-sensing, and self-healing. These special issues focus on the emerging ideas that permit the development of improved or new fiber-reinforced concrete composites and characterizations of the features of advanced fiber-reinforced concretes. Original research papers and authoritative review journals explain the present findings in the advanced fiber-reinforced concrete composite field are anticipated to cover a variety of topic. The potential topics include, but are not restricted to structural applications of advanced fiber-reinforced concrete composites, fiber-bridging behaviors, multiple micro-cracks, strain-hardening behaviors, property characterization, nanofiber reinforced concrete composites, advanced fiber-reinforced cement-free composites, ultra-high performance fiber-reinforced concretes, multifunctional fiber-reinforced concrete composites, and advanced fiber-reinforced concrete composites.

Keywords Nanofiber-reinforced composites · Structural applications · Micromechanical characterization

4.1 Introduction

Fiber-reinforced concretes are perfect for enhancing the mortar and concrete toughness and durability performance (Ohama 2011; Zajac et al. 2018; Wang et al. 2020). In concretes, fibers help reduce hazardous spalling at high temperatures, increase absorption of energy, increase strength, and reduce shrinkage cracks. There are various reasons for adding fibers in concretes. The homogenous distributions in the concretes is one of the major fiber benefits. Other benefits are minimize costs and save time in the process of construction, completely or partly substitute conventional

N. Makul, *Principles of Cement and Concrete Composites*, Structural Integrity 18,
https://doi.org/10.1007/978-3-030-69602-3_4

reinforcing steel, improve abrasion resistance and toughness, improve the explosive spalling resistance, increase ductility and load capacity, increase shear and flexural strength, control and minimize sizes because of early-age shrinkages, and better fresh concrete cohesion.

4.1.1 Definition and Characteristic

Fiber-reinforced concretes (FRCs) are concretes that contain fibrous materials as shown typically in Fig. 4.1 that increase their structural integrity. They contain short discrete fibers that are randomly oriented and uniformly distributed. Fibers include natural, synthetic, glass, and steel fibers. In shotcretes, fiber-reinforcements are majorly applied (Jone and Grasley 2011; Ohama 2011; Wyrzykowski and Lura 2013). Also, fiber-reinforcement can be utilized in ordinary concretes. Fiber-reinforced ordinary concretes are majorly applied for pavements and on-ground floors. While still improving the tensile strengths many times, concretes reinforced with fibers are cheaper than traditional reinforcement.

In addition, the fiber-reinforced concrete properties change with varying densities, orientation, distribution, geometries, fiber materials, and concrete (Kene et al. 2012; Materazzi et al. 2013; Hakim et al. 2020). They minimize water bleeding and also reduce concrete permeability. Some of fiber types produce higher shatter resistance and impact abrasion in concretes.

4.1.1.1 Characteristic of Fiber-Reinforced Concrete

The fiber in fiber reinforced concrete adds some strength to the concrete by mechanically reinforcing the concrete (Jone and Grasley 2011; Kawashima and Shah 2011;

Fig. 4.1 Typical fiber for reinforcing cement-based composites

Ohama 2011). When one tries to break fiber-reinforced concrete, the fibers tend to make it hard so it does not just fall apart in big chunks. The fibers are meshed together, and the concrete has to pull away from the fiber in order to break. The top view is regular concrete. In the past, typical fibers have been made of glass (fiberglass), carbon, polyester, or steel. A product is now available that is called ultra-high performance fiber-reinforced concretes (UHPCs). This product uses very small diameter brass-coated steel fibers (Wyrzykowski and Lura 2013; Zajac et al. 2018). The steel fibers align themselves to reinforce the concrete and give it strength in tension.

In addition, the concrete that is used is a very special concrete (Ulm et al. 2010; Wyrzykowski and Lura 2013; Zajac et al. 2018). They have taken the coarse aggregate out of it and have a special mix that brings the strength, with the fibers, up to about 150 MPa in 28 to 56 days, and 95 MPa in just a few days. The use of fiber-reinforced concrete can reduce the length of splices and can reduce the amount of required reinforcement. The other cool thing about this UHPC is that it is watertight. When placed correctly, it will seal the joints (Jone and Grasley 2011; Wyrzykowski and Lura 2013; Novak and Kohoutkova 2018; Hakim et al. 2020). The drawback of this UHPC is that it is very expensive.

FRC is commonly used for shotcrete for mine and tunnel lining, as well as for precast segmental tunnel lining (Wyrzykowski and Lura 2013; Hakim et al. 2020; Szajerski et al. 2020). One significant reason it is used is that the fibers impart excellent post-crack performance, referred to as "toughness." The ground through which those structures pass may move over time, and the concrete is likely to crack. We need the concrete to maintain a good deal of strength, even after the concrete has cracked. As an aside, it is becoming more common to include synthetic microfibers in tunnel lining concrete for fire purposes. In the event of a fire, the microfibers melt and act as micro-sized safety valves (Jones and Grasley 2011; Han et al. 2014; Novak and Kohoutkova 2018). The molten plastic can be pushed out of the way to allow release of steam, rather than causing spalling of the concrete.

4.1.2 Classification

Under development in research labs or in practical application, a great diversity of different fiber-reinforced concretes can be found today. These include concretes with materials with very different structures and compositions and greatly different material characteristics (di Prisco et al. 2013; Demis et al. 2014; Guo 2014; Han et al. 2014; D'Alessandro et al. 2016). Generally, the tensile responses of fiber-reinforced concrete composites can be categorized in two distinctive classifications depending on the behavior after original cracking, namely either strain softening or strain hardening.

4.1.2.1 Mechanical Classification: Strain Hardening and Tension Softening

Including polymeric materials, wood, fiber-reinforced concrete, aluminum, and steel, many building materials show similar behaviors under mechanical loading (Kawashima and Shan 2011; Kene et al. 2012; Klyuev et al. 2018). It is important to realize that three scenarios can be accomplished when dealing fiber-reinforced concrete aspects: Both the softening and the hardening regimes are important enough to be put into consideration in structural designs, tension softening responses are significant enough that they can be put into consideration in structural contexts, and tension-softening responses are so important that they can be permitted to be considered in structural designs.

One can differentiate between deflection-softening and deflection-hardening behaviors within the strain-softening classification (Ulm et al. 2010; Ohama 2011; Otero-Chans et al. 2018). To examine the features of mostly strain-softening FRC composites through bending tests, several standards tests such as RILEM, ASTM, and JCI are available. But no standard tests are presently available to characterize strain-hardening responses in tensions. Such composites have been defined as HPFRC or FRC composites.

4.2 Manufacturing Process and Its Related Procedures

Fiber-reinforced concretes are produced by including fibers in cement matrices in a dispersed manner including: at least one plasticizer type, water, second aggregate particle with a maximum diameter of particles small equal to or small than 0.425 mm, a fineness modulus within a range of 0.4 to 0.8 and an average diameter of particles within a range of 0.1–0.3 mm; first aggregate particles having a maximum diameter of particles equal to or smaller than 2.5 mm, a fineness modulus within a range of 1.5–3.5, and an average diameter of particles within a range of 0.4–0.8 mm; particles of second pozzolanic reactants having activities lower than the particles of first pozzolanic reactants; and particles of first pozzolanic reactant particles having higher activities.

4.2.1 Designing Concept

To determine the fiber-reinforced concrete element performance, the properties of materials such as the residual strengths and its load-carrying capacities determined by standard tests—ASTM C1609—are inserted into the equations. Plain concrete tensile strength is negligible, therefore, not taken into account in the designs of traditional fiber-concrete sections (Ohama 2011; Wu et al. 2018; Wang et al. 2020). In the design process, effective tensile strengths of fiber-reinforced concrete sections are applied. The residual tensile strengths are obtained from the measured flexural strengths by conversion factor means because direct tensile tests on fiber-reinforced concrete are hard.

4.2.2 Materials Preparation

Rapid hardening fiber-reinforced concretes and high calcium sulfoaluminate and alumina in addition to fiber have been utilized for minimizing the time of setting and improving early foamed concrete strengths (Ohama 2011; Wetzel and Middendorf 2019; Wang et al. 2020). Fiber-reinforced concretes have been utilized in the ranges of 10–50% and 30–70%, respectively, as substitutes of cement to minimize hydration heat, minimize costs, and improve mix consistence while contributing toward long-term strengths. By cement mass, silica fumes up to ten percent have added to improve the cement strength. Alternate fine aggregate, viz. Lytag fines and expanded polystyrenes, quarry finer and foundry sand, reclaimed glasses, incinerator bottom ashes, crushed concretes and chalks, lime, and fly ashes were utilized either to utilize recycled/waste materials or minimize the foam concrete density (Ohama 2011; Kene et al. 2012; Hakim et al. 2020). Using lightweight coarse aggregates in foamed cement matrices, concretes with densities between 800 and 1200 kg/m^3 have been generated.

The water requirements for mixes depend on the use and constituent of admixtures and are controlled by the mix stability and consistency (Materazzi et al. 2013; Demis et al. 2014; Guo 2014; Lura and Terrasi 2014). The mix is too stiff at lower content of water causing bubbles to break. Higher contents of water make the mixes too thin to hold the bubbles resulting in bubbles' separation from the mixes and hence segregations. Water–cement ratios applied range from 0.400 to 1.250. Though superplasticizer is also at times used, in foamed concretes, its use can be potential reasons for foam instability. Therefore, admixtures' compatibility with foamed concretes is of significance. In the dosage ranges of 1–3 kg/m^3, chopped polypropylene fibers of 12 mm lengths are identified to improve the shear behaviors of foam concretes similar to those of ordinary concrete beams. The fiber usage is also noted to mitigate brittleness, while minimizing its costs and weights (Ohama 2011; Zajac et al. 2018; Wang et al. 2020). Optimal combinations of costs, workability, density, ductility, and strength may be acquired by choosing appropriate types of fibers, water–cement ratios of base mortars, and air contents.

4.2.3 Mixing, Transporting, Pouring, Compacting, and Curing

Transporting is carried out in first step (Akcay and Tasdemir 2010; Ulm et al. 2010; Wyrzykowski and Lura 2013; Otero-Chans et al. 2018). Stripping is involved in the second stage. In stripping, rocks are cleared and leveled using tools like bulldozer. At this stage, charges incurred include indirect costs, fuel costs, and labor costs. The blasting processes in the third phase. Blasting machinery is applied in the third phase. In this stage, the expenses include labor charges, fixed indirect costs, energy charges, machinery costs, costs of working capital, and capital costs. Stockpiling is the forth phase. It involves labor costs at approximated rates of $18/h (Ulm et al. 2010; Novak and Kohoutkova 2018; Zajac et al. 2018; Wetzel and Middendorf 2019). The firth phase is sorting in which tools such as excavator are allocated. The firth stage involves costs like costs of equipment maintenance, equipment charges, labor charges, fixed indirect costs, operating costs, costs of working capital, and capital costs. The sixth process is the crushing processes. Crushing processes involve magnetic separations, primary grinding, and secondary grinding. Tools applied include primary grinders, secondary grinders, and the shapers. In additions, cost of equipment maintenances, fuel charges, fixed indirect costs, labor expenses, and capita costs should be considered. The only benefit acquired in this processes is costs of maintenances that is easily saved in comparison with the processes of recycling because of the wears and tears of the machine blades deployed. Air sitting, screening, and washing processes are involved in the sixth phase. It involves recycled fuels and water in settling the dusts and other substances down (Ulm et al. 2010; Ohama 2011; Pereira et al. 2012; Song et al. 2018). The product phase ultimately involved finished commodity of various lengths sold at different amounts from $15.00 to $25.00.

Scholars (Karakurt et al. 2010) used two-phase mixing approaches to acquire best-quality the advanced fiber-reinforced concretes. The scholars utilized fibers treated with pozzolanic powders to enhance the advanced fiber-reinforced concretes properties. Also, other researchers (Harbulakova et al. 2014) designed two-phase mixing techniques for fostering high-grade applications of the advanced fiber-reinforced concretes. The investigators revealed that compared to conventional concretes, the 100% replacements of conventional concrete aggregates are possible by their mixing approaches to generate the advanced fiber-reinforced concretes with suitable features, although the optimum scenario takes place with 20% conventional concrete substitution. The results of strengths and slumps recommended that the new mixing methods contributed greatly to accomplish high flexural and compressive strengths and best workability. Besides, researchers (Karakurt et al. 2010) argued that the advanced fiber-reinforced concretes interfacial transition zones were realized by scanning electron microscopes (SEMs). The SEM findings established that the new techniques of mixings contributed to dense micro-structures. The inner bleedings in additions can be minimized by the methods of mixings.

With high cement contents, the advanced fiber-reinforced concretes are reported to have high carbonation resistances. Greater advanced fiber-reinforced concrete cement contents accomplish preferred compressive strengths. The advanced fiber-reinforced concrete tensile resistances increase with improved cement contents in concretes (Ohama 2011).

The decrease in the advanced fiber-reinforced concrete performances is associated with the water-to-cement ratios utilized in mix designs. As compared to the conventional concretes, advanced fiber-reinforced concretes need greater cement contents and lesser water-to-cement ratios to accomplish specific compressive strengths (Materazzi et al. 2013; Song et al. 2018; Zajac et al. 2018). The conventional concrete resistance to thawing and freezing at water-to-cement ratios of 0.290 was exceptionally high. But for advanced fiber-reinforced concrete, the same water-to-cement ratios failed to offer suitable freezing–thawing resistances.

Using a prolonged curing in wet environments is another approach of enhancing the advanced fiber-reinforced concrete performances (Kawashima and Shah 2011; Kene et al. 2012; Guo 2014; Klyuv et al. 2018). One of the most applied approaches to reduce the advanced fiber-reinforced concrete carbonation rate is extended curing. In advanced fiber-reinforced concrete, the carbonation depths are virtually two times lesser when the concretes are cured with water. More positive effects on advanced fiber-reinforced concrete are generated by the outer environmental curing condition than on conventional concretes. The differences in the splitting tensile strengths among the conventional concretes and advanced fiber-reinforced concrete are high when they are cured in the external environments (Materazzi et al. 2013; Song et al. 2018; Hakim et al. 2020). Besides, the water-cured advanced fiber-reinforced concrete carbonation depth is virtually double than those of concretes of air-cured advanced fiber-reinforced concrete. The reductions in the carbonation depth generated by water-curing could be partly because of greater concrete internal humidity.

4.2.4 Controlling Properties and Factors Affecting

The advanced fiber-reinforced concrete features are affected by various main aspects like air entrainments, cement contents, curing conditions, the advanced fiber-reinforced concrete moisture conditions, the qualities of the parent advanced fiber-reinforced concrete, the advanced fiber-reinforced concrete physical features, the advanced fiber-reinforced concrete size and type, the content of advanced fiber-reinforced concrete, and the water-to-cement ratios (Wyrzykowski and Lura 2013; Wetzel and Middendorf 2019). The aspects are explained below.

The advanced fiber-reinforced concrete physical characteristics substantially affect the hardened and fresh concrete properties. For example, the mixes of fresh concretes with coarse-textured and angular particles become harsh and thus are hard to complete (Harbulakova et al. 2014; Novak and Kohoutkova 2018; Song et al. 2018). Also, the concrete finishability and workability can be influenced because of the high advanced fiber-reinforced concrete absorption. Besides, the greater advanced fiber-reinforced concrete pore volumes can influence transportation features (permeability and water absorption), strengths, and the porosities of the concretes.

The toughened and fresh features of concretes are highly affected by the contents of advanced fiber-reinforced concretes utilized as a full or partial conventional concrete replacements. Using seven independent variables (Kene et al. 2012; Materazzi et al. 2013; Wetzel and Middendorf 2019) created a model regarding the amount and type of aggregates to predict the performances of concretes for 0.0–100% conventional concrete replacements with advanced fiber-reinforced concrete. The researchers reported that the concretes generated with advanced fiber-reinforced concrete had lesser elastic modulus and compressive strengths than traditional concretes. The greater advanced fiber-reinforced concrete contents also increase water absorptions but decrease densities, therefore resulting in increased concrete porousness. The applications of 50–100% rough advanced fiber-reinforced concrete decrease concrete density and increase the absorption of water by roughly 2.11–3.50% and 0.14–0.38%, respectively. Besides, scholars (Li and Yang 2017; Proske et al. 2018; Han et al. 2019) noted that with increase grounded advanced fiber-reinforced concrete contents, the resistances to penetrations of chloride ions, whereas concrete tensile splitting and compressive strengths reduced. Also, the investigators that concrete drying shrinkages improved with increases in advanced fiber-reinforced concrete contents. It can cover be regulated by lowering water-to-cement ratios.

Researchers (Jone and Grasley 2011; Ohama 2011; D'Alessandro et al. 2016) explore the influences of quality of parent concretes on the advanced fiber-reinforced concrete features. The scholars reported that the advanced fiber-reinforced concrete water absorptions increase with the improved parent concrete strength. This is because the concretes with greater strengths fundamentally need higher cement contents, therefore increasing the mortar amount sticking to the aggregate. The mix water content adjustments therefore are necessary for the ne concretes, including the advanced fiber-reinforced concrete obtained from old concretes of greater strengths to derive the preferred workability. The porous advanced fiber-reinforced concrete affects the advanced fiber-reinforced concrete strengths (Ulm et al. 2010; Novak and Kohoutkova 2018; Zajac et al. 2018). The proportional loss in the tensile or compressive strengths of new concretes because of the advanced fiber-reinforced concrete use is more substantial when it is obtained from weaker old concretes than strong old concretes.

Scholars (Novak and Kohoutkova 2018; Otero-Chans et al. 2018; Szajerski et al. 2020) applied three diverse sizes of aggregates in assessing the advanced fiber-reinforced concrete size influences on the concrete properties. The greater reductions in the elasticity modulus were derived for the concretes prepared with tinier advanced fiber-reinforced concrete sizes. On the other hand, they reported that the strengths increase with increases in the optimum RCA sizes. Also, they found that the concrete water absorptions reduce with the increases in the maximum advanced fiber-reinforced concrete sizes (Ulm et al. 2010; di Prisco et al. 2013; Wyrzykowski and Lura 2013; Klyuev et al. 2018). This is because of the comparatively lower contents of weak mortars stuck to large-sized aggregate.

The aggregate moisture conditions affect the concrete workability. The concrete original slumps (workability measurements) are greatly reliant on the original free water contents of the concrete mixes. Researchers (Ulm et al. 2010; Materazzi et al. 2013; Demis et al. 2014; Hakim et al. 2020) revealed that while the air-dry and saturated surface-dry advanced fiber-reinforced concrete shows ordinary initial slumps and slump losses, the oven-dry advanced fiber-reinforced concrete results in faster slump losses and greater original slumps. With high cement contents, the advanced fiber-reinforced concretes are reported to have high carbonation resistances. The advanced fiber-reinforced concrete tensile resistances increase with improved cement contents in concretes. Through appropriate air entrainments, durable advanced fiber-reinforced concrete can be generated. The air entrainments are as successful for conventional concretes as for advanced fiber-reinforced concrete concretes (Ulm et al. 2010; Wyrzykowski and Lura 2013; Zajac et al. 2018). Besides, entrained air application is more successful than reducing the water-to-cement ratios to enhance advanced fiber-reinforced concrete resistances to thawing and freezing.

4.2.4.1 Fresh State

The advanced fiber-reinforced concrete can affect the fresh concrete characteristics because of their higher porosities, absorptions, surface coarseness, and angularity (Lai et al. 2010; Harbulakova et al. 2014; Lura and Terrasi 2014; Li and Yang 2017). As realized from the available literature, the advanced fiber-reinforced concrete effects on the major concrete fresh features are provided and explained thereafter. Scholars (Karakurt et al. 2010) noted that the fiber-reinforced concrete strengths may be boosted by either allowing reclaimed aggregates to soak up parts of mixing water without or with pozzolanic liquids during mixings or soaking the fiber-reinforced concrete in mixes of pozzolanic liquids such as colloidal silicas or water before mixing of concretes. The micro-cracks in fiber-reinforced concrete are expected to be filled up by the absorbed pozzolanic liquids or absorbed water with cement gels during pozzolanic reactions or cement hydrations (Karakurt et al. 2010). Therefore, the fiber-reinforced concrete strengths can be increased. Another scholar (Akcay and Tasdemir 2010) reported that the correct water adjustment to cement ratios for the mixes of concretes might enhance the advanced fiber-reinforced concrete strengths. Other scholars (Beushausen et al. 2014) suggested higher advanced fiber-reinforced concrete cement contents and lesser water-to-cement ratios than those of parent concretes to accomplish similar compressive strengths.

4.2.4.2 Plastic State

The concrete elasticity modulus or plastic state is increased by an aggregate of greater elastic modulus. The elasticity modulus or plastic state of concrete thus reduces with fiber content increase in concretes. Typically, the concrete elasticity modulus having advanced fiber-reinforced concrete is 10–33% lesser than those of conventional concretes. Researcher (Materazzi et al. 2013) illustrated that the applications of 30% fibers in concretes generated about 15% elastic modulus reductions. As compared to conventional concretes, the advanced fiber-reinforced concrete elastic modulus nevertheless can be lesser up to 46%. Concrete elasticity modulus reductions are because of the fact that advanced fiber-reinforced concrete generally have lesser elastic modulus as compared to conventional concretes. Also, the concrete elastic modulus reductions are related to the improved overall contents of mortars (recycled and new) that have lesser elastic moduli than most natural aggregate concretes.

4.2.4.3 Early-Age State

Generally, the advanced fiber-reinforced concrete flexural strengths are lesser than those of conventional concrete strengths (Pan et al. 2018). The three-day advanced fiber-reinforced concrete flexural strengths were greater than those conventional concretes. At the period of twenty-eight days, the strength was lower. In their research, the conventional concretes increased strengths progressively and had greater flexural strengths than the concretes of advanced fiber-reinforced concrete at later ages. Advanced fiber-reinforced concrete produced any notable adverse effects on the concrete flexural strengths (Kene et al. 2012; Pan et al. 2018; Wetzel and Middendorf 2019). However, with adequate strengths, advanced fiber-reinforced concrete can be generated for various uses, at times even with 100% substitutions of conventional concrete.

4.2.4.4 Hardened State

The concrete bond strengths are indications of the interconnecting features of pastes and aggregates. The coarse advanced fiber-reinforced concrete surfaces lead to better bond than conventional concretes. For testing the bonds between reinforcements and concretes that included 100, 50, and 0% advanced fiber-reinforced concrete, researchers (Novak and Kohoutkova 2018) used 150 mm × Ø100 cylindrical samples with fixed mild and ribbed reinforcements (diameter = 12.00 mm and embedded lengths = 150 mm). The findings indicated that the bonds between reinforcements and advanced fiber-reinforced concrete concretes are greatly affected by the fiber inclusions in concretes. The findings disclosed that the advanced fiber-reinforced concrete bond strengths were 10–20% greater than those of traditional concretes. Such contradicting findings mean that more studies are needed to examine the fiber effects on the concrete bond strength.

Mechanical Properties

The concrete mechanical features rely on the concrete mechanical characteristics. It was established that advanced fiber-reinforced concrete mechanical features are lower as compared to the mechanical features of conventional concrete aggregates from the available literature (Karakurt et al. 2010; Jones and Grasley 2011; Harbulakova et al. 2014). The advanced fiber-reinforced concrete strengths may be boosted by either allowing reclaimed aggregates to soak up parts of mixing water without or with pozzolanic liquids during mixings or soaking the advanced fiber-reinforced concrete in mixes of pozzolanic liquids such as colloidal silicas or water before mixing of concretes (Materazzi et al. 2013). The micro-cracks in advanced fiber-reinforced concrete are expected to be filled up by the absorbed pozzolanic liquids or absorbed water with cement gels during pozzolanic reactions or cement hydrations (Jones and Grasley 2011). Therefore, the advanced fiber-reinforced concrete strengths can be increased.

When the material loss because of wear becomes higher, a greater AVV is obtained. Usually, the advanced fiber-reinforced concrete aggregate abrasion value is greater than that of NCAs. The classic abrasion values vary from 20.0 to 45.0%, which is greater than the values of conventional concrete (Kawashima and Shah 2011; Kene et al. 2012; Guo 2014; Klyuev et al. 2018). Despite of their origins, the advanced fiber-reinforced concrete abrasion values are basically below the acceptable optimum limits for structural applications (fifty percent total weight). The aggregate effect values (AEVs) indicate the aggregate resistances to dynamic loads. The conventional concrete impact values (15–20%) are lower than that of advanced fiber-reinforced concrete (20–25%) (Kawashima and Shah 2011; Kene et al. 2012; Guo 2014; Klyuev et al. 2018). The attached cement and mortar pastes make concrete less strong and thus lead to a higher aggregate impact values for advanced fiber-reinforced concrete. The aggregate grounded values (AGVs) indicate measures for aggregate resistances to grinding under progressively applied compressive loads (Kawashima and Shah 2011; Kene et al. 2012; Guo 2014; Klyuev et al. 2018). It has been established that AGVs of convectional concretes (14–22%) is substantially lesser than that of advanced fiber-reinforced concrete (20–30%) from the literature and as indicated in table two. This is anticipated due to the comparatively weak mortars and cement pastes fixed to the particles of advanced fiber-reinforced concrete.

Durability

Through appropriate air entrainments, durable advanced fiber-reinforced concrete can be generated. The air entrainments are as successful for traditional concretes as for advanced fiber-reinforced concrete (Kens et al. 2012; Pan et al. 2018; Wetzel and Middendorf 2019). Besides, entrained air application is more successful than reducing the water-to-cement ratios to enhance advanced fiber-reinforced concrete resistances to thawing and freezing. The advanced fiber-reinforced concrete durability and strength can be enhanced by utilized appropriate pozzolanic substances (Karakurt et al. 2010; Jones and Grasley 2011; Harbulakova et al. 2014). Scholars (Jones and Grasley 2011) indicated that the applications of 65% fibers and 30% pulverized fly ashes improved the advanced fiber-reinforced concrete compressive strengths to the control concrete cast levels with virgin granite gravels. Other scholars (Jones and Grasley 2011) indicated that the fibers and pulverized fuel ashes were successful in increasing the chloride ion penetration resistances into advanced fiber-reinforced concrete.

4.3 Repair and Maintenance

All structures deteriorate over time. They need regular maintenance and cleaning and occasional repairs and repainting to stay in tiptop shape. It is amazing how fast a structure will show the effects of neglect. New office or apartment buildings will start looking aged and ugly if not well maintained for a period of time. Without traffic, even a roadway that has been abandoned will crack up and weeds will quickly take over. Drains will become clogged, concrete will stain, paint will weather, dirt and grime will accumulate, joints will start to leak, dry rot will affect wood members, iron will rust, and the sun, rain, and wind will take their toll. An abandoned house generally appears to deteriorate much faster than one that is lived in.

One of the major problems with such structures worldwide is that the owners often scrimp on maintenance due to the costs. Even in the USA, much of the infrastructure has deteriorated badly due to lack of routine maintenance. Bridges and roads deteriorate. Reservoirs fill up with sediment. Weeds take over. Pipelines leak or break. Termites do their damage. Sealants crack. Construction materials degrade. Differential settlement causes damage. In time, weeds and forest reclaim the land.

4.3.1 Assessment Principles

Various blends with water and superplasticizer variations are utilized as trial judges of rendering the testing robust (Karakurt et al. 2010; Ohama 2011; Chen et al. 2018; Otero-Chans et al. 2018). Various rheological features, workability properties, and mechanical characteristics are the parameters to be tested. The strength of different blends having variations of advanced fiber-reinforced concrete aggregate is tried for robustness. The applications of high-level recycled aggregate have been determined to adversely influence the concrete characteristics (Ulm et al. 2010; Kene et al. 2012; Song et al. 2018; Wetzel and Middendorf 2019). The elastic static and strength modulus reduce with the increases in the advanced fiber-reinforced concrete aggregate contents.

The applications of the high-performance of recycled materials have been determined to adversely influence the concrete features (Jones and Grasley 2011; di Prisco et al. 2013; Chen et al. 2018; Klyuev et al. 2018). The elastic static modulus and the compressive strengths reduce with the increases in the recycled material contents. Nevertheless, the reductions can be sufficiently compensated by the applications of lo w/b ratio. Besides, cement additions in the advanced fiber-reinforced concrete aggregate at the same replacements and w/b ratio reduce the tensile splitting strengths and the compressive strengths. Nevertheless, partial fly ash applications as cement replacements decrease the RCA shrinkage (Lai et al. 2010; Li and Yang 2017; Klyuev et al. 2018). The fly ashes result in the long-term and greater advanced fiber-reinforced concrete aggregate durability because of pozzolanic fly ash reactions.

It is vital to identify the amounts saved in recycling to compute the benefits and costs of advanced fiber-reinforced concrete aggregate materials (Ulm et al. 2010; Materazzi et al. 2013; Otero-Chans et al. 2018; Proske et al. 2018). Also, extra amounts are saved because there are no production and no dumping of new concretes. Nevertheless, important data about costs is still not available. The agencies should carry out detailed debates with reps from various construction and building firms to establish this and determine the social and environmental expenses of utilizing advanced fiber-reinforced concrete aggregate in comparisons to conventional materials. The researcher analyzed traditional concrete approaches to determine the costs of utilizing advanced fiber-reinforced concrete aggregate against conventional materials (Qiu et al. 2013; Lura and Terrasi 2014). The building and construction wastes need to be put into consideration in the traditional methods in the initial phase of the cost analyses.

4.3.2 Repair and Strengthening

The concrete properties and mix proportions are influenced by the physical advanced fiber-reinforced concrete characteristics (Ohama 2011; Zajac et al 2018; Wetzel and Middendorf 2019). The fundamental properties like absorptions, pore volumes, bulk density, specific gravity, textures, and shapes of advanced fiber-reinforced concrete are overall inferior than those of conventional concrete aggregate because of the impurities and residual cement mortar/paste presence (Ohama 2011; Yang et al. 2017; Wang et al. 2020). The magnitudes of the impacts differ with quantities and nature of fiber cement mortars or pastes that are present in advanced fiber-reinforced concrete. The advanced fiber-reinforced concrete can affect the fresh concrete characteristics because of their higher porosities, absorptions, surface coarseness, and angularity (Kene et al. 2012; Wetzel and Middendorf 2019). As realized from the available literature, the advanced fiber-reinforced concrete effects on the major concrete fresh features.

The higher coarseness and angularity particles of advanced fiber-reinforced concrete reduce the concrete workability and make them more hard to complete appropriately (Jone and Grasley 2011; Demis et al. 2014; Otero-Chans et al. 2018; Zajac et al. 2018). The reduction levels in workability increase with the improved advanced fiber-reinforced concrete proportion in the mixes of the concretes. More water is thus needed for advanced fiber-reinforced concrete to get similar conventional concrete workability. The mixes of concretes integrating advanced fiber-reinforced concrete generally fulfill the original requirements of slumps. Nevertheless, the greater advanced fiber-reinforced concrete absorptions can result in a rapid workability loss, therefore, restricting the period needed for putting and completing of concretes (Ohama 2011; Otero-Chans et al. 2018; Wu et al. 2018). The challenges related to the rapid workability loss should be tackled by modifying and managing the advanced fiber-reinforced concrete water content before mixings, not by addition of more water at the worksites.

Generally, the conventional concrete bleeding is greater than that of advanced fiber-reinforced concrete (Ulm et al. 2010; Otero-Chans et al. 2018; Song et al. 2018). During mixing, some of the ancient cement pastes rubbed off advanced fiber-reinforced concrete and generate extra fines in the mixes of concretes. These fines therefore minimize the concrete bleeding after adsorbing some of the mix water. At lower free water contents, the higher quantities of fines also increase the concrete mix cohesiveness. Besides, the increased surface coarseness and angularity at greater advanced fiber-reinforced concrete contents contribute to the improvements of concrete cohesiveness. The concrete mix stability enhances because of improved cohesiveness and minimized bleeding (Otero-Chans et al. 2018; Zajac et al. 2018). Therefore, the advanced fiber-reinforced concrete segregation resistance can be compared to that of traditional concrete aggregates.

The advanced fiber-reinforced concrete effects on the toughened concrete characteristics may be substantial or insignificant relying on their physical features, gradations, contents, types, and sources. The toughened advanced fiber-reinforced concrete characteristics decrease with the conventional concrete substation level by advanced fiber-reinforced concrete (Kawashima and Shah 2011; Pan et al. 2018; Hakim et al. 2020). Without greatly affecting the hardened concrete properties, up to 30% of conventional concrete by advanced fiber-reinforced concrete on weight basis as a general principle. As noted from the available literature, the variety of transformations in the toughened concrete features because of advanced fiber-reinforced concrete.

The compressive advanced fiber-reinforced concrete strengths are often lesser than those of conventional concretes (Kene et al. 2012; Demis et al. 2014; Pan et al. 2018; Zajac et al. 2018). The compressive advanced fiber-reinforced concrete strengths most commonly are 10–20% lesser than those of conventional concretes. Depending on the advanced fiber-reinforced concrete quality, it may also be reduced up to 20%. Normally, the greater air contents found in the mixes of concretes having advanced fiber-reinforced concretes can also result in lesser values of strength. Advanced fiber-reinforced concrete nevertheless can have the same and at times greater compressive strengths than conventional concretes if the advanced fiber-reinforced concrete are obtained from old concrete sources that are initially generated with lesser water-to-cement ratios than the new concretes (Materazzi et al. 2013; Demis et al. 2014; Yang et al. 2017; Song et al. 2018; Hakim et al. 2020).

By weights, advanced fiber-reinforced concrete generated positive effects on the concrete compressive strengths up to the replacement degree of 10%. For the advanced fiber-reinforced concrete content, the compressive strength is improved by more than 30% (Ohama 2011; Materazzi et al. 2013; Wetzel and Middedorf 2019). The concrete compressive strengths are much higher when advanced fiber-reinforced concretes are utilized in the oven-dry states. 20–30% increments in compressive strengths are realized because of the advanced fiber-reinforced concrete applications in the case of high-performance concretes.

Also, the course advanced fiber-reinforced concrete can influence the concrete compressive strengths (Karakurt et al. 2010; Chen et al. 2018; Zajac et al. 2018; Wetzel and Middendorf 2019). The advanced fiber-reinforced concrete compressive strength is affected by the rough concrete to fine concrete ratios of the advanced fiber-reinforced concrete source concretes. The lesser coarse to fine aggregate ratios result in greater mortar quantities attached to particles of coarse advanced fiber-reinforced concrete and therefore leads to increment in the advanced fiber-reinforced concrete strengths. These increments are even higher when advanced fiber-reinforced concrete fines are utilized. Thus, the fine advanced fiber-reinforced concrete uses in concretes are generally recommended.

Regarding the fiber-reinforced concrete effects on the concrete splitting tensile strengths, adequate literate is available. Fiber-reinforced concrete splitting tensile strengths are higher than those of conventional concretes (Novak and Kohoutkova 2018; Wetzel and Middendorf 2019; Szajerski et al. 2020; Wang et al. 2020). Various investigators reported that the fiber-reinforced concrete splitting tensile strengths are 10–20% higher than those of conventional concretes. During the time of three months to one year, no statistically notable reductions in tensile strengths took place. Fiber-reinforced concrete generates greater tensile strengths than conventional concretes.

The fiber-reinforced concrete flexural strengths are greater than those of conventional concrete strengths. Typically, the fiber-reinforced concrete flexural strengths are greater than those of conventional concretes. The seven-day fiber-reinforced concrete flexural strengths are greater than those concretes of conventional concretes (Ulm et al. 2010; Mo et al. 2012). At the period of one month, the strength is even higher. The fiber-reinforced concrete increased strengths progressively and had greater flexural strengths than the conventional concretes at later ages. Fiber-reinforced concrete produced notable adverse effects on the concrete flexural strengths. With adequate strengths, fiber-reinforced concrete can be generated for various uses, at times even with conventional strengths.

The concrete permeability relies on both the concrete matrix permeability (binder and the cement pastes) and the included aggregate absorption capacities. Also, the concrete permeability is influenced by its continuities of the pores, distributions, and size and porosity. For equivalent water-to-cement ratios, the fiber-reinforced concrete permeability seems to be greater than those of conventional concretes. The fiber-reinforced concrete permeability may be 100–450% greater than those of conventional concretes.

4.3.3 Demolition and Decommissioning

Demolitions of ancient buildings and building of new structures are regular idea because of natural disasters, expansions of traffic directions, city rearrangement, structural deteriorations, and purpose change (Ulm et al. 2010; Wetzel and Middendorf 2019; Wang et al. 2020). In the European Union, around 850 tons of demolition and construction wastes are produced every year. This represents about 30% of the overall waste production. The trash generated from construction demolitions alone is about 123 tons yearly the U.S. A massive concrete wastes are derived from the demolitions of ancient concrete buildings. These concrete wastes are most commonly disposed to landfills, therefore causing substantial health hazards and environmental loads. Besides, the increasing landfill charges and land shortages worsen these ecological problems. The concrete waste utilization in sustainable developments can alleviate such challenges. Over thirty year ago, research studies began to explore the advanced fiber-reinforced concrete characteristics (Jones and Grasley 2011; Kawashima and Shah 2011; Hakim et al. 2020). In the past, most studies carried out were mainly restricted to the production of non-structural grade advanced fiber-reinforced concrete because of undesirable physical advanced fiber-reinforced concrete properties, like high absorptions of water that increase the demand of water for specific workability.

Currently, concrete is the most basic construction material (Lai et al. 2010; Kene et al. 2012; Klyuev et al. 2018). The production of concrete uses up a lot of resources mainly sand where its harvesting has a great environmental impact on riverbeds and sand borrow pits hence the need to innovate methods to supplement concrete production with other substances like advanced fiber-reinforced concrete from demolished sites and heap of waste mortar in mega projects blended with laterites soil as a sand substitute. In the recent past, concrete structures have been undergoing demolition resulting in the debris which commonly has been utilized as landfills. The environmental sustainability is a key focus, and the need to keep construction costs low, the debris can be crushed to achieve the fine recycled concrete aggregates (FRCAs) which can be utilized in replacing the fine natural aggregates (sand) partially. This can be used to produce fresh concrete, therefore, achieving a sustainable construction industry. In the recent past, concrete structures have been undergoing demolition resulting in the debris which commonly has been utilized as landfills partially (Materazzi et al. 2013; Novak and Kohoutkova 2018). This can be used to produce fresh concrete, therefore, achieving a sustainable construction sector.

As the aggregate may be retrieved from concrete residuals after placing is finished, waste generation needs not to be a challenge within the construction. Usually, this represents the concrete sludge and 2.00% of the aggregate fresh material remaining after the reclamation of the aggregates. The remaining concrete sludge can be mixed with the liquid from washing of the truck. The slurry of cement pastes can be dehydrated, desiccated, grounded, and utilized to substitute NCAs with fine RCAs (Karakurt et al. 2010; Jones and Grasley 2011; Kawashima and Shah 2011). The only sure way of minimizing construction costs is the utilization of the available local resources and adoption of innovative construction materials. Applying RCA and blended laterites as fine aggregates on all these accounts in concretes appear to the best, particularly in the place in which the mixed substances are freely available other than the natural concrete aggregates such as river sand.

4.4 Practical Applications and Implementation

Structural grade concretes can also be generated by the applications of advanced fiber-reinforced concrete. The advanced fiber-reinforced concrete utilization in high-performance as well as high-strengthen structural concrete is possible through correct quality control and mix design and with the addition of silica fumes. Nevertheless, more priority has been given to illustrate the advanced fiber-reinforced concrete effects on material durability and hardened and fresh characteristics. The significant findings of this study is that advanced fiber-reinforced concrete production setup can be utilized in large-scale production of recycled materials at low costs. Also, front-loader type and overhead bin kind of industries may be applied to reduce the advanced fiber-reinforced concrete incremental costs. Nevertheless, the pricing effects and demand and supply aspects of recycled materials pose different challenges that are rarely accounted for.

4.5 Recycling and Other Applications

Recycled concrete aggregates (RCAs are utilized as recycled concrete wastes in the present structures of green concretes. The RCAs are sustainable concrete wastes that in the long-run can substitute the demands for natural aggregates, processes that would in turn result to their preservations. Concrete wastes are collected and grounded using recycling procedures to generate grinded concretes that are then utilized in structural concretes in which they substitute natural aggregates that is rough in this process. Presently, sustainable developments are major issues throughout the world. The sustainable development concept has now become a guiding standard for construction and building sector across the globe. Reuse and recycling of concrete wastes can be a successful strategy to accomplish sustainability in the construction sector. Many administrations worldwide have introduced different controls with the goals of minimizing the use of virgin aggregate and improving the concrete waste recycling for reuse as materials whenever it environmentally, technically, and economically acceptable. Nevertheless, it has been noted that the majority of concrete industries have been reluctant to produce RCA and utilize it to its full potentials. Manufacturing plants are yet to embrace the use of RCA not only because of its unclear concrete performances but also because of its unexplored production industry processes that are however to be determined. The most concrete plants have been observed to reluctant in the RCA production and usage in its optimum potentials. Plants are yet to embrace production of RCAs not only because of their unclear substance performances but also because its unexplored operations of production plants that are yet to be determined.

4.6 Conclusions

Various researchers have studied advanced fiber-reinforced concrete features. Advanced fiber-reinforced concrete is an effective way to reduce constructions and demolition wastes. However, the properties of concrete with advanced fiber-reinforced concrete have to be investigated. Scholar indicated that the concrete strength could be reduced up to 40% when advanced fiber-reinforced concrete was used. However, various elements such as replacement level, moisture content, type of RCAs and water-to-cement ratio affect the advanced fiber-reinforced concrete strength. Different from conventional concretes, advanced fiber-reinforced concrete has typically higher water absorption, which could significantly affect concrete properties, especially the workability. Researchers demonstrated that concrete with drier advanced fiber-reinforced concrete had a more considerable slump and faster slump loss than that of saturated advanced fiber-reinforced concrete. It is clear that the strengths for the entire fine aggregate replacement increase with time with the blends having higher percentages of the combination gaining at a higher percentage. This depicted the characteristics of the presence of lateritic material which from the literature review available, it stated that the strength for the material increase with the structure's life.

References

Akcay B, Tasdemir MA (2010) Effects of distribution of lightweight aggregates on internal curing of concrete. Cem Concr Compos 32(8):611–616

Beushausen H, Gillmer M, Alexander M (2014) The influence of superabsorbent polymers on strength and durability properties of blended cement mortars. Cem Concr Compos 52:73–80

Chen Q et al (2018) A stochastic micromechanical model for fiber-reinforced concrete using maximum entropy principle. Acta Mechanica 229(7):2719–2735

D'Alessandro A et al (2016) Investigations on scalable fabrication procedures for self-sensing carbon nanotube cement-matrix composites for SHM applications. Cem Concr Compos 65:200–213

Demis S, Efstathiou M, Papadakis V (2014) Computer-aided modeling of concrete service life. Cem Concr Compos 47:9–18

di Prisco M, Matteo C, Daniele D (2013) Fibre-reinforced concrete in fib Model Code 2010: principles, models and test validation. Struct Concr 14(4): 342–361

Guo Z (2014) Principles of reinforced concrete. Butterworth-Heinemann

Hakim II, Putra N, Agustin PD (2020) Measurement of PCM-concrete composites thermal properties for energy conservation in building material. AIP Conf Proc. 2255(1). AIP Publishing LLC

Han B et al (2014) Nanotip-induced ultrahigh pressure-sensitive composites: principles, properties and applications. Compos Part A Appl Sci Manuf 59:105–114

Han B et al (2019) Nano-engineered cementitious composites: principles and practices. Springer

Harbulakova VO et al (2014) Different aggressive media influence related to selected characteristics of concrete composites investigation. Int J Energ Environ Eng 5(2–3):82

Jones CA, Grasley ZC (2011) Short-term creep of cement paste during nanoindentation. Cem Concr Compos 33(1):12–18

Karakurt C, Kurama H, Topcu IB (2010) Utilization of natural zeolite in aerated concrete production. Cem Concr Compos 32(1):1–8

Kawashima S, Shah SP (2011) Early-age autogenous and drying shrinkage behavior of cellulose fiber-reinforced cementitious materials. Cem Concr Compos 33(2):201–208

Kene KS, Vairagade VS, Sathawane S (2012). Experimental study on behavior of steel and glass fiber reinforced concrete composites. Bonfring Int J Ind Eng Manag Sci 2(4):125–130

Klyuev SV et al (2018) The fiber-reinforced concrete constructions experimental research. Mater Sci Forum 931. Trans Tech Publications Ltd

Lai WL et al (2010) Characterization of the deterioration of externally bonded CFRP-concrete composites using quantitative infrared thermography. Cem Concr Compos 32(9):740–746

Li J, Yang E-H (2017) Macroscopic and microstructural properties of engineered cementitious composites incorporating recycled concrete fines. Cem Concr Compos 78:33–42

Lura P, Terrasi GP (2014) Reduction of fire spalling in high-performance concrete by means of superabsorbent polymers and polypropylene fibers: small scale fire tests of carbon fiber reinforced plastic-prestressed self-compacting concrete. Cem Concr Compos 49:36–42

Materazzi AL, Ubertini F, D'Alessandro A (2013) Carbon nanotube cement-based transducers for dynamic sensing of strain. Cem Concr Compos 37:2–11

Mo L, Deng M, Wang A (2012) Effects of MgO-based expansive additive on compensating the shrinkage of cement paste under non-wet curing conditions. Cem Concr Compos 34(3):377–383

Murtazaev SY et al (2018) Impact of technogenic raw materials on the properties of high-quality concrete composites. In: International symposium engineering and earth sciences: applied and fundamental research (ISEES 2018). Atlantis Press

Novak J, Kohoutkova A (2018) Mechanical properties of concrete composites subject to elevated temperature. Fire Saf J 95:66–76

Ohama Y (2011) Concrete-polymer composites—the past, present and future. Key Eng Mater 466. Trans Tech Publications Ltd

Otero-Chans D et al (2018) Experimental analysis of glued-in steel plates used as shear connectors in Timber-concrete-composites. Eng Struct 170:1–10

Pan X et al (2018) Effect of inorganic surface treatment on surface hardness and carbonation of cement-based materials. Cem Concr Compos 90:218–224

Pereira EB, Fischer G, Barros JA (2012) Effect of hybrid fiber reinforcement on the cracking process in fiber reinforced cementitious composites. Cem Concr Compos 34(10):1114–1123

Pourjavadi A et al (2013) Interactions between superabsorbent polymers and cement-based composites incorporating colloidal silica nanoparticles. Cem Concr Compos 37:196–204

Proske T et al (2018) Concretes made of efficient multi-composite cements with slag and limestone. Cem Concr Compos 89:107–119

Qiu Y et al (2013) Review on composite structural health monitoring based on fiber Bragg grating sensing principle. J Shanghai Jiaotong Univ (Science) 18(2):129–139

Snoeck D, Pel L, De Belie N (2018) Superabsorbent polymers to mitigate plastic drying shrinkage in a cement paste as studied by NMR. Cem Concr Compos 93:54–62

Song Q et al (2018) Steel fibre content and interconnection induced electrochemical corrosion of ultra-high performance fibre reinforced concrete (UHPFRC). Cem Concr Compos 94:191–200

Szajerski P et al (2020) Radiation induced strength enhancement of sulfur polymer concrete composites based on waste and residue fillers. J Clean Prod 271:122563

Ulm F-J et al (2010) Does microstructure matter for statistical nanoindentation techniques? Cem Concr Compos 32(1):92–99

Wang X et al (2020) Interfacial characteristics of nano-engineered concrete composites. Constr Build Mater 259:119803

Wetzel A, Middendorf B (2019) Influence of silica fume on properties of fresh and hardened ultra-high performance concrete based on alkali-activated slag. Cem Concr Compos 100:53–59

Wu C-R et al (2018) Improving the properties of recycled concrete aggregate with bio-deposition approach. Cem Concr Compos 94:248–254

Wyrzykowski M, Lura P (2013) Controlling the coefficient of thermal expansion of cementitious materials—a new application for superabsorbent polymers. Cem Concr Compos 35(1):49–58
Yang Z et al (2017) In-situ X-ray computed tomography characterisation of 3D fracture evolution and image-based numerical homogenisation of concrete. Cem Concr Compos 75:74–83
Zajac M et al (2018) Impact of microstructure on the performance of composite cements: why higher total porosity can result in higher strength. Cem Concr Compos 90:178–192

Chapter 5
Principles of Reactive Powder Concrete

Abstract There have been multiple developments and improvements in concrete technologies that had impacts on structural system in recent years. The cement mixtures are used with silica fumes in place of steel fibers and silica fumes (pozzolanic materials) in the form of superplasticizers and are classified as ultra-high-strength concretes to enhance the slab structural behaviors. This chapter explains current study literature associating with reactive powder concrete (RPC). So far, there are no available official design codes on RPC. Current information associated with the present topic can be classified according to the historical background and development of RPC and its mechanical properties and durability.

Keywords Reactive powder concrete · High-performance concretes · Mechanical properties · Durability

5.1 Introduction

Reactive powder concretes (RPCs) are emerging composite materials that shall enable the concrete sector to construct structures that are sensitive to environment, durable, and strong, improve use of materials, and produce economic gains. Comparison of the durability, physical, and mechanical characteristics of high-performance concretes (HPCs) and reactive powder concretes indicates that reactive powder concretes possess lesser permeability and stronger flexural and compressive strengths in comparison with high-performance concretes. The work reviewed the available information on reactive powder concretes and presented the laboratory results and investigations comparing RPCs with HPCs. Also, potential applications and specifics of RPC have also been discussed. High-performance concretes (HPCs) are not just simple mixtures of aggregates, cements, and water. They contain chemical admixtures and mineral components with specific properties that provide unique characteristics to the concretes. The developments of HPCs results from materializations of the applications of advanced scientific equipment, a new science of concretes, and a new science of admixtures to monitor the microstructures of concretes.

In the early 1990s, reactive powder concretes (RPCs) were discovered in France. In July, The Canadian Sherbrook Bridge, the first reactive powder concrete structure

N. Makul, *Principles of Cement and Concrete Composites*, Structural Integrity 18,
https://doi.org/10.1007/978-3-030-69602-3_5

in the universe, was built. Reactive powder concretes (RPCs) are defined as high ductility and ultra-high-strength cementitious composites with enhanced physical and mechanical features. They are unique concretes that are enhanced by microstructures via correct gradients of all substances in the mixes to generate maximum density. The highest strength hydrates are created by optimization of Portland cement chemical characteristics and pozzolanic characteristics of silica fumes or fine materials.

Substantial advancements have been accomplished to design various types of concretes in the last decades. Compared to the familiar, technology characterizes the RPCs as concretes types of technical revolutions high-strength concretes (HSC) and steel fiber concrete (SFC) are special kinds of improved concretes. The most important factor in high-performance and steel fiber concretes is the improved strength. These materials that lately available demonstrate the strength characteristics and high-improved durability compared to high-performance or familiar concretes. They are classified as ultra-high-performance concretes (UHPCs). As described by Blais and Couture (1999), the RPCs are the most significant enhanced "high-tech" materials. Types of these new cement-built composite materials of ultra-high-strength strengths and superior improvements are defined as RPCs. In the durability matrices and RPC mixtures, a specific quantity of short steel fibers can be utilized to overcome the high instability difficulty and improve the reactive powder concrete. RPCs have been rapidly applied in many areas of constructions such as mining engineering, high rise buildings, and bridge erection (Dauriac 1997).

The high-performance concrete (HPC) requirements utilized for Indian Nuclear Power Industries' nuclear waste containment structures are high durability, good workability, uniform densities, moderate, E value, and normal compressive strengths (Basu 1999). There needs to analyze reactive powder concretes concerning their durability and strengths to recommend their applications for atomic waste containment structures.

The three commercial names of ultra-high-performance concrete (UHPC) or RPC can be summarized as shown in Fig. 5.1 (Theresa et al. 2008). Ductal is produced in France through Lafarge Company having a compressive strength in the variety of 100–200 MPa. A similar kind of concrete is called commercially Béton Special Industrial (BSI) concrete. This is a self-setting, fiber-reinforced concrete that was offered by the firm Eiffage construction in association with Sika. Direct tensile strength of this concrete ranges from 8 to 10 MPa for 28 days' age obtained, and the characteristic compressive strength is found to be 150 MPa (Brandt 2008).

A third UHPC called Béton Composite Vicat (BCV) is industrialized by the cement manufacturer Vicat and Vinci group Resplendino (2004). The concept of

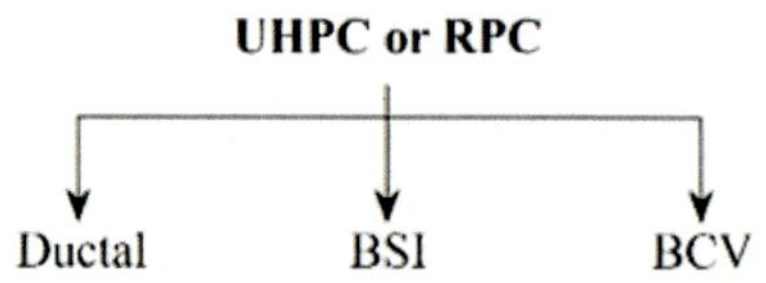

Fig. 5.1 Schematic diagram showing types of RPC (Theresa et al. 2008)

RPC is followed on the basis that materials with a minimum damage or defect, like small cracks, internal spaces which will have better load carrying capacity and larger capacity to withstand.

Structural designers continually pursue new construction methods and ideas that will make their buildings beautiful, attractive, functionally effective, and cost efficient. Historically, the improvement in the structures depended strongly upon the characteristics of engineering materials. A new kind of material with excellent properties usually results in a revolution in structures and constructions. These are as well true for concrete and steel structures. Without designing new design criteria for RPC structures, the structures cannot however be developed with new concretes such as reactive powder concretes. With regular design reference that non-conventional resources do not deal fine, the criteria for RPC structures must be carried out in making application of these materials. Because the present codes restrict extra strengths, concrete compressive strengths that are higher than 80 MPa may not be possible in several nations. Thus, it is necessary to examine the mechanical properties of columns, slabs, and beams of RPCs.

The following standards for enhancing RPCs were proposed by Richard and Cheyrezy (1995).

- Improvements of homogeneities by eliminating coarse concretes from the mixes.
- Employments of pozzolanic properties of silica fumes.
- Developments and improvements of densities of compacted concretes by the granular mixture optimizations.
- Improve the pressure workability before and throughout setting to create compactions and reductions in water–cement ratios by the optimum applications of superplasticizers.
- After settings via the temperatures, treatments shall result to microstructure improvements.
- Ductility enhancements by adding small-sized steel fibers.

5.2 RPC Compositions

Reactive powder concretes are composed of superplasticizers, steel fibers (optional), and very fine powder such as silica fumes, quartz powders, quartz, sand, and cements. Applied at their optimal dosages, the superplasticizers decrease the water-to-cement ratio (w/c) while enhancing the concrete workability. Very dense matrices are accomplished by optimization of the dry fine powder granular packing. The durability and ultra-high strengths of reactive powder concretes give this compactness. Reactive powder concrete has strength varying from 200 to 800 MPa. Structural designers continually pursue new construction methods and ideas that will make their buildings beautiful, attractive, functionally effective, and cost efficient. Historically, the improvement in the structures depended strongly upon the characteristics of engineering materials. A new kind of material with excellent properties usually results in a revolution in structures and constructions. These are as well true for concrete and steel structures. Without designing new design criteria for RPC structures, the structures cannot however be developed with new concretes such as reactive powder concretes. With regular design reference that non-conventional resources do not deal fine, the criteria for RPC structures must be carried out in making application of these materials. Because the present codes restrict extra strengths, concrete compressive strengths that are higher than 80 MPa may not be possible in several nations. Thus, it is necessary to examine the mechanical properties of columns, slabs, and beams of reactive powder concretes.

Together with recommendations on how to accomplish RPC performance, Table 5.1 outlines RPCs' salient characteristics. The different RPC ingredients and their choice considerations are described in Table 5.2. The RPC mixture designs chiefly involve the dense granular skeleton development. The granular mixture optimizations can be accomplished either by software of particle size distribution like LISA (designed by Elkem ASA material) or by the applications of packing models. Experimental methods have been preferred this far for mixture designs of reactive powder concretes. Different RPC mixture proportions obtained from available information are presented in Table 5.3 (Richard and Cheyrezy 1994; Blais and Couture 1999; Matte and Moranville 1999; Staquet and Espion 2000).

The water demands (water quantity for minimum concrete flow) are the main parameters that decide the mixture quality. In fact, the mixture void indices are associated with the entrapped air and the sum of water demands. Optimum contents of water are evaluated utilizing the parameter relative densities (d_0/d_S) after choosing mixture designs according to minimum demand of water. The d_0 or d_S here represents the mixture compacted density (no air or water) and the concrete density, respectively.

Table 5.1 RPCs characteristics improving strengths and homogeneity

RPCs characteristics	Descriptions	Proposed values	Failure types eliminated
Reductions in aggregates to-matrix ratios	Limitations of sand contents	The paste volumes are at least 20% higher than the non-compacted sand void indices	Through any external sources such as formworks
Reductions in the sizes of aggregates	In reductions in the coarsest aggregate sizes by factors of about 50, coarse aggregate is substituted by fine sands	Optimum fine sand size is 600 μm	Thermo-mechanical and chemical and mechanical
Improved mechanical properties	Enhanced paste mechanical characteristics by adding of silica fumes	Modulus value of Young in a range of 50–75 GPa	The mechanical stress field disturbance

Table 5.2 RPC component selection parameters

Component	Selection parameter	Functions	Particle sizes	Type
Superplasticizers	Less hindering characteristics	Reduction of w/c	–	Polyacrylate-based
Sands	Low costs and readily available Good hardness	Aggregatesgive strengths	150–600 μm	Crushed and natural
Cements	C_4AF: 7.4%; C_3A: 3.8%; C_2S: 22%; C_3S: 60%; (optimum)	Primary hydrate production Binding materials	1–100 μm	Medium fineness OPC
Silica fumes	Very less impurity quantities	Secondary hydrate production Filling the voids Improve rheologies	0.1–1 μm	Procured from ferrosilicon industries (greatly refined)
Steel fibers	Good aspect ratios	Improve ductility	L: 13–25 mm Ø: 0.15–0.2 mm	Straights

Table 5.3 Typical mixture proportions of RPC

Materials	Richard and Cheyrezy (1995)				Matte and Moranville (1999)	Staquet and Espion (2000)
	Non-fibered		12 mm fibers		Fibered	Fibered
Portland cement	1	1	1	1	1	1
Silica fume	0.25	0.23	0.25	0.23	0.325	0.324
Sand	1.1	1.1	1.1	1.1	1.43	1.43
Quartz powder	–	0.39	–	0.39	0.30	0.30
Superplasticizer	0.016	0.019	0.016	0.019	0.018	0.021
Steel fifer	–	–	0.175	0.175	0.275	0.218
Water	0.15	0.17	0.17	0.19	0.20	0.23
Compacting pressure	–	–	–	–	–	–
Heat treatment temperature (°C)	20	90	20	90	90	90

Relative densities show the concrete packing levels, and their optimum value is one. The mixture designs need to be such that the packing densities are maximized.

The high RPC strengths make RPCs highly brittle. Generally, the steel fibers are included to improve the ductility of RPC. Typically, with diameters of 0.15 mm, straight steel fibers are about 13 mm long. At 3% by volume and ratio of between

1.5, the fibers are introduced into mixtures. The cost-efficient maximum dosage is similar to ratios of about 155 kg/m^3 or 2% by volume.

5.3 Mechanical Properties and Durability of RPC

RPC first construction goes to the Richard and Cheyrezy (1994, 1995) when the first papers were available in 1994 and 1995, respectively. According to these papers, two classes of RPC were manufactured with target compressive strength of 200 and 800 MPa. It was recommended that the size of maximum aggregate in UHPC is less than 600 μm.

Aitcin (1998) defined the concrete as "high strength based on its compressive strength measured at a specified age, but with the progress in research studies, the concrete which called high-strength is now named high performance concrete and this is due to not only to its strength but because: it gives enhanced performance such as durability and corrosion resistance." Great strides have been made in the developments of types of higher strengths in the form of high-performance concretes (HPCs), ultra-high-performance concretes (UHPCs), very high-performance concretes (VHPCs).

Jungwirth (2002) investigated the influences of heat treatment specimens of reactive powder concretes at 90 °C in comparison with the samples that are water cured at 20 °C on splitting and compressive strengths of tensile concretes. The sampled findings and results indicated an increase for splitting tensile strengths vary from 25 to 35 and for compressive strengths range from 120 to 180 MPa when the heat treatments were accomplished at 90 °C. The mixture of reactive powder concretes contained 24.7 kg/m^3 of superplasticizers, water/cement ratios of 0.240, 158 kg/m^3 steel fibers with l/d 26/0.160, 240 kg/m^3 silica fumes, 1086 kg/m^3 furnace sands, and 638 kg/m^3 of cements.

Silvia et al. (2003) investigated the effects of various types of fibers on the behaviors of reactive powder concretes. The mixtures of RPC were made from 12.30 kg/m^3 of superplasticizer, water/cement ratio of 0.24, 157 kg/m^3 steel fibers with l/d 25/0.160, 225 kg/m^3 silica fumes, 944.00 kg/m^3 sands and size of particles from 0.1 to 1.0 mm, and 905.00 kg/m^3 of cements. The mixtures were also made of 180 kg/m^3 of steel fiber. The different types of fibers used include deformed galvanized steel fiber (30/0.62), deformed steels (30/0.62), deformed steels (30/0.45), and brass-plated steel (13/0.18). When the brass-plated fibers were used in the RPC mixtures, the results indicated higher flexural and compressive strengths than RPC having other different types of fibers.

Using the test of two-point flexural tensile strength, Voo et al. (2003) studied RPC specimens. With a span of 400 mm, the prism attained sections of 100 mm squares. The steel fiber type used in the prism test consisted of either 30-mm end-hooked fiber or 13 mm straight fiber filled with 2% concrete volume. The flexural strengths are 25.2 MPa for end-hooked fiber and 23.2 MPa for straight fiber. The researchers found out there was an increase in the strength of RPC by over 8% for end-hooked fiber associated with straight fiber.

Al-Wahili (2005) studied the mechanical properties of RPC properties of ash in addition to rice husk. The concrete samples were under compression and exposed to various stages of thermal curing when it was hardened. The mixture consists of ordinary Portland cement, very fine sand (less than 600 μm), and rice husk ash. In this investigation, a dose of 10, 15, and 20% of rice husk by cement weight, steel fibers, and high range water decreases the admixture at optimum amount of dosage. Results demonstrated which was probable to create a concrete having compressive strength equal 132 MPa, flexural strength equal to 19.1 MPa, dynamic modulus equal to 48.61 GPa. The water absorption was 0.3 kg/m^2 in 15 days.

Jeffery et al. (2006) studied the structural behavior and performance of five normal and synthetic fiber-reinforced concrete slabs of dimensions (2.2 m^3 × 2.2 m^3 × 0.15 m^3) m under flexural loading. Two volumes of fraction fibers (0.32 and 0.48%) were used. Results specified that the addition of fibers do not alter the first crack load of plain concrete slabs, but the flexural cracking load of normal concrete increases by 25 and 32% and the ultimate load capacity of normal concrete increases by 20 and 34% with the addition of fibers by volume of fraction of 0.32 and 0.48%.

Ali (2006) aimed in part of his work in modifying reactive powder concrete by adding some styrene butadiene rubber (SBR) polymers and coarse aggregates to replace the high reactivity metakaolin (HRM) and the superplasticizers, as active mineral powders to be utilized as alternatives to the silica fumes. The natural aggregates of maximum sizes of 10 mm were used to replace part of the binders and fine aggregates. While flexural, compressive, and splitting strengths hand a little reduction, this resulted in an increase in water/cement ratio and cement/high reactivity metakaolin ratio.

Thomas and Ramaswamy (2007) investigated the impact of adding fibers on mechanical characteristics of concrete. The models were derived through established 60 data of test results by the regression analysis for different mechanical characteristics of steel fiber-reinforced concrete. Characteristics of the different strengths are the cubes and the cylinder compressive strengths, the tensile strengths (modulus of ruptures), modulus of elasticity, post-cracking performances, Poisson's ratio, and the strain at peak compressive stress results. The strength of reinforced concretes which contains steel fibers was related using the recommended prototypes with the test data obtained from the study and with the various other test data cited in the literature. The variable which is considered is strength of concrete, that is, 35, 65 and 85 MPa for normal strength, moderately high-strength, high-strength concrete, respectively, and steel ratios contents are 0, 0.5, 1.0 and 1.5%. The recommended mode predicted the test significantly and data quite accurately. The research specifies that the contact of the fiber matrix involved to the improvement and development of mechanical characteristic resulting from the outline of fibers that differs from current models and combinations based on the mixtures.

Husain (2008) modified an original reactive powder concrete (ORPC) which included a superplasticizer cement mixed and contained silica fume and fine steel fibers by using local metal admixture as high reactivity metakaolin as a replacement for silica fume, the steel fibrous was substituted with polypropylene fibers (PPF), and the part of mechanical characteristics improved the compressive and flexural strengths of MRPC and was enhanced punching shear and flexural strengths of normal and reinforced concrete flat plates. It was found that addition of PPF has increased the mechanical properties, where the cracks were more accurate and larger amount than normal and slabs reinforced did not contain fibers. The research results showed increase in ultimate load of rupture for non-reinforced MRPC slabs (without fibers) by 50% while for the reinforced MRPC slabs (without fibers) by about 17% in punching shear strengths, where the strength of concrete increases by 11.6%.

Also, the punching shear strengths are enlarged by about 5–46% compared to reinforced concrete slabs discluding fibers; while the enlargement in flexural strengths alternated between 6.1 and 41%, the ultimate load of fiber-reinforced concrete slabs was greater than panels discluding fibers.

Ibraheem (2008) presented an experimental research to found the widespread compressive stress–strain relations below the difference in pozzolanic material type, silica fume, metakaolin. The use of high volume of fraction of steel fibers and silica fume in the mix of RPC increases the compressive strengths, durability, ductility, and densities of the concretes and decreases its absorptions. To develop equations for determining the nominal flexural moment capability of RPC-reinforced rectangular units, an ideal compressive stress block for RPC sections was recommended and used under the bending moment. Also, the results of the experimental research showed that reactive powder concrete mixtures with the silica fume provide the maximum value of compressive strength, density, and less rate of absorption as compared to compressive strength ranges from 164 to 195 MPa.

An ideal steel fiber-reinforced concrete (SFRC) model was presented and designed by Fernando et al. (2009) to be applied in elements of plate. The elements were acknowledged to the cracks. The load systems and fiber contents were considered as variables in this study. The investigation utilized three different load systems and four different fiber contents. Three plates were tested for each combination, providing a complete of 36 units as shown in Fig. 5.2. The load systems were as indicated in figure two and the fiber contents were 90, 70, 50 and 0 kg of fiber/m^3 of concretes. The finite model elements portrayed good relationships regarding the experimental. The model is capable of simulating the answers in the cracking processes, involving a stress increase after cracking and a fast stress decline. The additions of fibers do not characterize suitable improvements to the concrete compression strength developments.

Shahidan (2009) studied the influence of using steel fibers on the performance of laminated reinforced concrete slabs through computer simulations. In this investigation, the imitation of limited elements for the analysis of ordinary reinforced concrete slabs and steel fiber-reinforced concrete slab due to different percentage of 1, 1.5, 2, 2.5 and 3% of fibers. The loading was applied by gradual increase of every 2 kN up to collapse. The analysis result indicated that using steel fibers in reinforced concrete will affect the concrete ductility, toughness, energy absorption, and concrete compressive strength. The aims of the research were to compare the

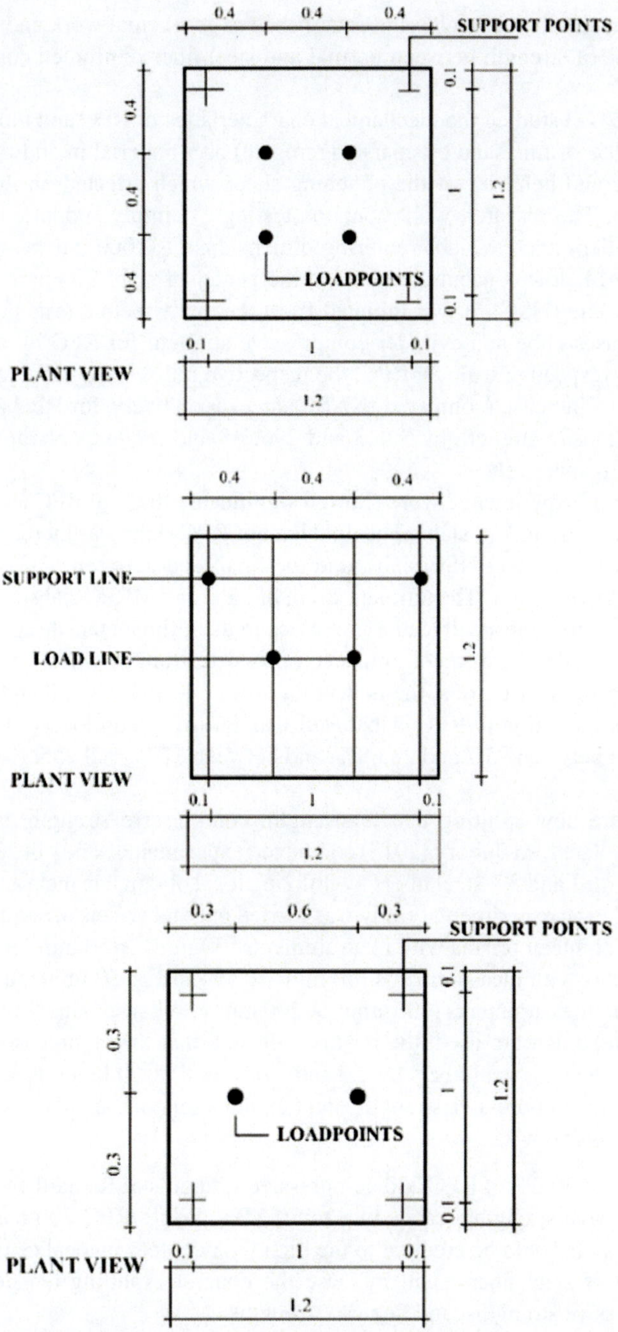

Fig. 5.2 Type of loading on plate elements (Fernando et al. 2009)

finite element simulation analysis in established experimental work and to compare the properties of strength between normal and steel fiber-reinforced concrete slabs (SFRC).

Hassan (2012) studied the mechanical characteristics of RPC and modified RPC (replacing 50% of fine sand by coarse aggregate) as a material in addition to study the influence and behavior of the punching shear which effected on the RPC and MRPC slabs. The research work contains testing of simply supported reinforced reactive powder concrete slabs requiring dimensions of (1000 mm × 1000 mm × 50 or 70 mm) below concentrated load at the center of slab. Compared to normal strength concrete (NSC), it was initiated from the experimental tests that using V_f by 2% increases cube and cylinder compressive strength for RPC by about 176.6 and 214.7%, respectively. For MRPC, the respective ratios were 155.5 and 180.8%. Percentages of increase (compared to NSC) are, respectively, for RPC and MRPC, for splitting tensile strength by 308.3 and 296.6% and modulus of rupture 405.54 and 334.6%, respectively.

The result of experimental work showed significant effects of RPC and MRPC on punching shear strength of slabs. The final load of RPC slabs was increased between 39.05 and 181.50% over NSC slabs, and the final load between 63.8 and 138.5% which is for MRPC slabs. The ultimate strain of RPC and MRPC slabs indicated that the increase in fiber ratio will lead to increases in the ultimate tensile strain between 15 and 99% and decreases the ultimate compressive strain between 1.5 and 33.3%. When adding of steel fibrous, the deflection at final load is significantly enlarged. The steel fiber addition to RPC slabs results in an increase in load in the center at ultimate load between 93.7 and 283.8%, and between 122.8 and 255.9% for MRPC slabs.

To measure how splitting tensile strength, compressive strength, and flexural strength are related, Al-Jubory (2013) conducted experimental study in 2013 on RPC reinforced with 1 and 2% steel fibers by utilizing local obtainable materials treated at 20 and 80 °C. The experimental study was carried out on two sets of samples. All the set involved eighteen prisms with measurements (50 mm^3 × 50 mm^3 × 300 mm^3), fifty-four cubes with measurements (50 mm^3 × 50 mm^3 × 50 mm^3), and eighteen cylinders with measurements (100 mm^2 × 200 mm^2). When about 2% of steel fibers with 20°C treatment are used, the results indicated that the compressive strengths are 74 MPas. Adding steel fibers by 1.0 and 2.0% shall result in increase in splitting tensile strength, compressive strength, and flexural strengths, and the results that are acquired are as follows:

1. Equaled to 30.0% of RPC 200 compressive strengths of Richard and Cheyrezy, compressive strengths reached up to 70 MPa. While RPC 200 has free C_3A, these results could be credited to the high C_3A content in local cements.
2. Adding of steel fibers shall increase the concrete splitting tensile strengths, compressive strengths, and flexural strengths.
3. At early ages, curing at high water temperatures of 80°C increased the compressive strengths as compared with curing at 20°C and reduced it at one month.

4. Qasim (2013) tested 31 specimens (21 one-way slab and 10 of two-way slabs. Several important parameters were chosen to study their effects on experimental, finite element, and analytical program for behavior and performance of reinforced RPC slabs with openings. These parameters are five steel fiber contents, three silica fume contents, three reinforcement ratios, two opening locations, four sizes and three shapes, grade of concrete and elements number in terms of load deflection response, moment–curvature response, failure load, strain in concrete, strain in steel, and cracking pattern. In general, a good accord concerning the finite element solutions, analytical program, and experimental results was found.

Results of experimental, analytical, and finite element presented that the increases in steel fiber contents and silica fume lead to an increase in the ultimate load and thus lead to increase in the deflection at failure load and curvature, while more content of steel ratio may cause to greater load at failure and less content of both deflections at failure and curvature. Moreover, flexural toughness increased with the increase in fiber content and silica fume contents.

Majeed et al. (2015) studied (6) samples differed only in the void volume V_v and the volume of steel fibers (V_f) and at that moment use carbon fibers for support. Also, the aim of enhancement by fibers, carbon polymer (CFRP), was found out to be effective structural members. The results of prototypical tests that were rehabilitated by used carbon fiber (CFRP), compared with the same samples before strengthening and tested will lead to reduce the failure, increased extreme durability, with the increase ranging from 51.6 to 96.2%. When treated after the failure and the correct use in the happening of the failure in whole or in part in the ceiling, and tested by using the same way and condition where the slabs was examined through it, is knowing that ceilings was strengthened and functioned in the same way.

This study concluded it is feasible to use HCS for constructions as a floor and roofing system. It was shown also that these slabs are efficient after rehabilitation using carbon fiber (CFRP) strips.

Qaseem (2015) performed an experimental work of punching shear for six reduced scale-reinforced concrete slab specimens distributed into two group (square and trapezoidal slabs) behaviors. All specimens of slabs were prepared with square shape and having dimensions of 450 mm^2 × 450 mm^2 and thickness of 50 mm. While the trapezoidal slabs were made with dimensions 450 mm width, 620 mm length, 50 mm thickness, the upper side was made with 200 mm width.

Each group consisted of three specimens that were identical in size and shape but contained different percentages of steel fibers 0, 0.5 and 1% of total volume. The results showed that the punching shear strength increased by about 62.5 and 100% for square slabs and it is about 8.3 and 41.7% for trapezoidal slabs having 0.5 and 1% of steel fibers, respectively, as presented in Fig. 5.3.

Gholamhoseini et al. (2016) studied sixteen full-scale continuous slab samples having different kinds of bonding between the concrete slab and steel decking (e.g., greased, standard decking) with various types and quantities of reinforcement in concrete (such as steel fibers, mesh, or ordinary reinforcement). The slabs were

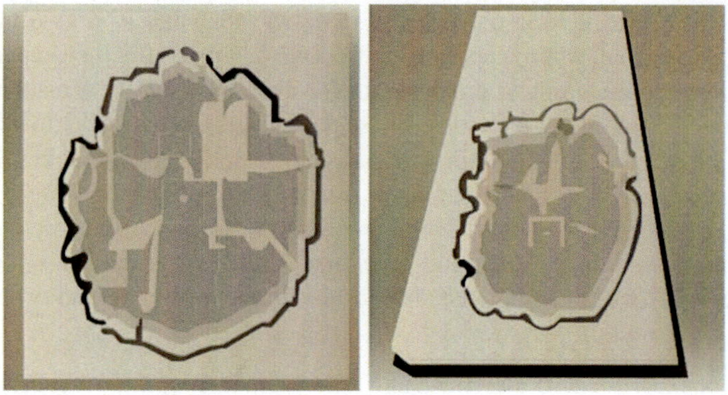

Fig. 5.3 Failure pattern of flat slab samples (Qaseem 2015)

6.3 m long, 1.2 m wide, and 0.15 m thick involving of two distances of 3.0 m plus a 150 mm addition from each external support in both long-term and short-term tests. Each slab was continuous on the inner support, and roller was supported at each end. The shrinkage of concrete was accounted for an age of 90 days where serviceability behavior of slabs was considered.

Based on the designated RPC 200 and RPC 800, the family of reactive powder concretes includes two concrete forms that provide interesting implicational potentials in various regions. Table 5.4 gives the mechanical characteristics for the two RPC types. The high RPC flexural strengths are because of the additions of steel fiber. Typical mechanical RPC characteristics in comparison with compressive strength 80MPa conventional HPC are indicated in Table 5.5 (Bickley and Mitchell 1999). Fracture toughness measures energy absorbed per material unit volume to fractures. RPC portrays high ductility and fracture toughness as compared to HPC. RPC has ultra-dense microstructures apart from its unique mechanical properties, giving beneficial durability and waterproofing properties. Therefore, these substances can be utilized for nuclear and industrial waste storage plants (Richard and Cheyrezy 1994).

Reactive powder concretes have ultra-high durability properties originating from their improved corrosion resistance, restricted shrinkage, low permeability, and extremely low porosities. There are no penetrations of gases or liquids through reactive powder concretes as compared to HPCs. Given in Table 5.6, the RPC properties

Table 5.4 Comparisons of reactive powder concrete 200 and reactive powder concrete 800

Properties	RPC 200	RPC 800
Flexural strengths (MPa)	30–60	45–141
Compressive strengths (using steel aggregates) (MPa)	–	650–810
Compressive strengths (using quartz sands) (MPa)	170–230	490–680
Heat-treating (°C)	20–90	250–400
Pre-setting pressurizations (MPa)	None	50

Table 5.5 Comparisons of high-performance concretes (HPCs) (80 MPas) and 200

Properties	Performance concretes (HPCs) (80 MPa)	Reactive powder concretes (RPCs) 200
Fracture toughness (J/m^2)	$<10^3$	$30*10^3$
Modulus of elasticity (GPa)	40	60
Flexural strengths (MPa)	7	40
Compressive strengths (MPa)	80	200

Table 5.6 Reactive powder concretes (RPCs) durability compared to high-performance concretes (HPCs)

Properties	Values
Chloride ions diffusions	25 times lesser
Rate of corrosions	8 times lesser
Water absorptions	7 times lesser
Abrasive wears	2.5 times lesser

allow their uses where physical wears significantly limit the life of other concretes and in chemically aggressive environments (Richard and Cheyrezy 1994).

5.4 Reactive Powder Concrete (RPC) Limitations

The least expensive traditional concrete elements are essentially substituted or eliminated by more expensive components in typical mixture designs of RPCs. The fine sands utilized in RPCs become similar to the traditional concrete fine aggregates. The Portland cements play the fine aggregate role. The silica fumes play the role of cements. The optimizations of the mineral elements alone lead to notable increases in cost above and over those of traditional concretes (five to ten times greater than of high-performance concretes (HPCs). RPC needs to be utilized in areas in which some of the notable properties of the materials can be completely used and where huge weight saving can be got (Aitcin 2000). In compression members in which durability concerns are at stake such as in marine conditions, reactive powder concretes (RPCs) can even substitute steels owing to their durability. The long-term characteristics are unknown because reactive powder concretes (RPCs) are in their developing phase.

5.5 Conclusions

Substantial advancements have been accomplished to design various types of concretes in the last decades. Compared to the familiar, technology characterizes the reactive powder concretes as concrete types of technical revolutions, high-strength concretes (HSCs), and the steel fiber concrete (SFC) are special kinds of improved concretes. The most important factor in high-performance and steel fiber concretes is the improved strength. These materials that are lately available demonstrate the strength characteristics and high-improved durability compared to high-performance or familiar concretes. They are classified as ultra-high-performance concretes (UHPCs). As described by Collepardi (2001), the reactive powder concretes are the most significant enhanced "high-tech" materials. Types of these new cement-built composite materials of ultra-high strength and superior improvements are defined as reactive powder concretes. In the durability matrices and RPC mixtures, a specific quantity of short steel fibers can be utilized to overcome the high instability difficulty and improve the reactive powder concrete. Reactive powder concretes have been rapidly applied in many areas of constructions such as mining engineering, high rise buildings, and bridge erection (Yao and Zhou 2005).

References

Aitcin PC (1998) The art and science of high performance concrete technical paper from the internet

Aitcin PC (2000) Cements of yesterday and today Concrete of tomorrow. Cem Concr Res 30:1349–1359

Ali AS (2006) Mechanical properties and durability of original and polymer modified reactive powder concrete exposed to oil products. Ph.D. thesis, Building and Construction Engineering Department, University of Technology, Baghdad, p 218

Al-Jubory N (2013) Mechanical properties of reactive powder concrete (RPC) with mineral admixture, vol 28(5). University of Mosul, pp 92–101

Al-Wahili A (2005) Mechanical properties of steel fiber reinforced reactive powder concrete. M.Sc. thesis, Building and Construction Engineering Department, University of Technology, Baghdad, p 123

Basu PC (1999) Performance requirements of HPC for Indian NPP structures. Indian Concr J 539–546

Bickley JA, Mitchell D (1999) A state-of-the-art review of high performance concrete structures built in Canada: 1990–2000. 96–102

Blais PY, Couture M (1999) Precast, prestressed pedestrian bridge—world's first reactive powder concrete structure. PCI J 44: 60–71

Brandt AM (2008) Fiber reinforced cement-based (FRC) composites after over 40 years of development in building and civil engineering. Compos Struct 86(1–3):3–9

Casadei P, Nanni A, Ibell T (2003) Experiments on two-way R.C. Slabs with openings strengthened with CFRP laminates. In: Proceedings of advancing with composites, American Concrete Institute, May 7–9 2003. ACI Structural Journal, Milano, Italy

Collepardi M (2001) Innovative concretes for civil engineering structures: SCC, HPC, and RPC, the new concrete

Danha LS, Khalil WI, al-Hassani HM (2015) Proposed model for uniaxial compression behavior of reactive powder concrete. J Babylon Univ Eng Sci 23(3):591–606

Dauriac C (1997) Special concrete may give steel stiff competition, building with concrete. Seattle Daily J Comm

Fernando LG, Domingoa A, Lázaroa C, Colinab CL, Pelufoa M (2009) Model for the behavior of steel fibers reinforced concrete in plates elements. In: Proceedings of the international association for shell and spatial structures (IASS) symposium, Valencia evolution and trends in design, analysis and construction of shell and spatial structures, 28 September–2 October 2009

Gholamhoseini A, Khanlou A, MacRae G, Scott A, Hicks S, Leon R (2016) An experimental study on strength and serviceability of reinforced and steel fibre reinforced concrete (SFRC) continuous composite slabs New Zealand Eng Struct 114(1):171–180. Department of Civil and Natural Resources Engineering, University of Canterbury, Christchurch 8140

Hassan H (2012) Punching shear behavior of normal and modified reactive powder concretes slabs. PhD. thesis, AL-Mustansiriya University College of Engineering, Civil Engineering Department, Baghdad, 2012 September, p 222

Hussain LF (2008) Punching shear and flexural strength of fibrous modified reactive powder concrete flat plates. M.Sc. thesis, College of Engineering, Al-Mustansiriya University, Baghdad, January 2008, p 130

Ibraheem S (2008) Stress-strain relationships of reactive powder concrete. Ph.D. thesis, Building and Construction Engineering Department, University of Technology, Baghdad, p 173

Jeffery RR, Salah A, David AL, Klaus R, Gegory RV (2006) Effect of synthetic fibers on structural behavior of concrete slabs on ground. ACI Mater J 103(1):3–10

Jungwirth J (2002) Under-spanned bridge structure in reactive powder concrete (RPC). Swiss Federal Institute of Technology, Structural Concrete Laboratory, pp 1–6

Majeed W, Jumaa M, Abed al-rahman M (2015) Structural behavior of fibrous reinforced concrete hollow core one-way slabs strengthening by C.F.R.P. Tikrit J Eng Sci 22(1):101–111

Matte V, Moranville M (1999) Durability of reactive powder composites: influence of silica fume on the leaching properties of very low water/binder pastes. Cem Concr Compos 21:1–9

Qaseem TA (2015) Punching shear behavior of reactive powder concrete slabs with different shape. In: Second Engineering Scientific, Conference-College of Engineering University of Diyala, 16–17 December 2015, pp 61–73

Qasim O (2013) Behavior of reinforced reactive powder concrete slabs with openings. Ph.D. thesis, AL-Nahrain University College of Engineering, Civil Engineering Department, Baghdad, April 2013, p 224

Redha M, Mahmood K, Dawood M (2017) Punching shear behavior of continuous bubbled reinforced reactive powder concrete slab. J Babylon Univ Eng Sci 25(1):236–250

Resplendino J (2004) First recommendations for ultra-high-performance concretes and examples of application. In: Proceedings of the international symposium on ultra high performance concrete, University of Kassel, Kassel, Germany, 13–15 September 2004, pp 79–90

Richard P, Cheyrezy M (1994) Reactive powder concrete with high ductility and 200–800 MPa compressive strength. In: ACI SP144-24, January 1994, pp 507–518

Richard P, Cheyrezy M (1995) Composition of reactive powder concretes. Cem Concr Res 25(7):1501–1511

Shahidan S (2009) Behavior of steel fiber reinforced concrete slab due to volume fraction of fiber. M.Sc. theses, University of Putra Malaysia, December 2009, p 69

Silvia C, Troli R, Monosi S, Orlando G (2003) The influence of the fiber type on the performance of RPC. Ind Italian Cem 786:334–341

Staquet S, Espion B (2000) Influence of cement and silica fume type on compressive strength of reactive powder concrete. In: 6th international symposium on HPC, University of Brussels, Belgium, pp 1–14

Theresa MA, Erron JP, Donald LM (2008) Ultra high-performance concrete for michigan bridges material performance phase-I. Center for Structural Durability, Michigan Technological Transpiration Institute, November 2008

Thomas J, Ramaswamy A (2007) Mechanical properties of steel fiber-reinforced concrete. J Mater Civil Eng 19(5):385–392

Voo YL, Foster SJ, Gilbert IR (2003) Shear strength of fiber reinforced reactive powder concrete girders without stirrups. In: UNICIV Report R-421, School of Civil and Environmental Engineering, The University of New South Wales, Sydney, Australia, 2052, 8

Yao ZX, Zhou J (2005) Study on fracture energy of reactive powder concrete reinforced by steel-polypropylene hybrid fiber. J Build Mater 8(4):356–360

Chapter 6
Principles of Tailor-Made Recycled Aggregate Concrete

Abstract Recycled concrete aggregates (RCAs) are utilized as recycled concrete wastes in the present structures of green concretes. The RCAs are sustainable concrete wastes that in the long run can substitute the demands for natural aggregates, processes that would in turn result to their preservations. The RCA use in concretes as full or partial substitutions of natural coarse aggregates (NCAs) is increasing interests in the building and construction sector because it minimizes the demands for natural aggregates. Besides, RCA utilization minimizes the adverse environmental impacts of virgin aggregate extractions from the natural resources and results in potential solutions to the ecological problems caused by concrete wastes. The aim of this paper is to critically review recycled concrete aggregates to produce ultra-performance concrete structures. This chapter provides a detailed review on RCAs application in concretes on the basis of the experimental information available in the published studies. In this chapter, the most significant chemical, physical, and mechanical features of recycled concrete aggregate are explained. Nevertheless, more focus has been given to explain the RCA effects on the concrete durability and fresh and hardened characteristics of the concrete. Also, this research identified the disparities existing in the current knowledge state on recycled concrete aggregates and natural concrete aggregates and offers some suggestions for future studies.

Keywords Durability · Hardened properties · Fresh properties · Recycled concrete aggregates · Natural coarse aggregates

6.1 Introduction

Concrete wastes are collected and grounded using recycling procedures to generate grinded concretes that are then utilized in structural concretes in which they substitute natural aggregates that is rough in this process (Abdel-Hay 2017). Presently, sustainable developments are major issues throughout the world. The sustainable development concept has now become a guiding standard for construction and building sector across the globe. Reuse and recycling of concrete wastes can be a successful strategy to accomplish sustainability in the construction sector. Many administrations worldwide have introduced different controls with the goals of minimizing

N. Makul, *Principles of Cement and Concrete Composites*, Structural Integrity 18,
https://doi.org/10.1007/978-3-030-69602-3_6

the use of virgin aggregate and improving the concrete waste recycling for reuse as materials, whenever it is environmentally, technically, and economically acceptable. Nevertheless, it has been noted that the majority of concrete industries have been reluctant to produce RCA and utilize it to its full potentials. Manufacturing plants are yet to embrace the use of RCA not only because of its unclear concrete performances but also because of its unexplored production industry processes that are however to be determined. The most concrete plants have been noticed to be hesitant in the RCA productions and usage in its optimum potentials (Abdel-Shafy and Mansour 2018). Plants are yet to embrace production of RCAs not only because of their unclear substance performances but also because its unexplored operations of production plants that are yet to be determined.

Demolitions of ancient buildings and building of new structures are regular idea because of natural disasters, expansions of traffic directions, city rearrangement, structural deteriorations, and purpose change (Al-Bayati, Tighe and Achebe, 2018). In the European Unions, around 850 tons of demolition and construction wastes are produced every year. This represents about 30% of the overall waste production. The trash generated from construction demolitions alone is about 123 tons per year of the U.S. massive concrete wastes are derived from the demolitions of ancient concrete buildings. These concrete wastes are most commonly disposed to landfills, therefore causing substantial health hazards and environmental loads. Besides, the increasing landfill charges and land shortages worsen these ecological problems. The concrete waste utilization in sustainable developments can alleviate such challenges (Abdulmatin et al. 2019). Over thirty years ago, research studies began to explore the TRAC characteristics. In the past, most studies carried out were mainly restricted to the production of non-structural rating TRAC concretes because of detrimental physical TRAC features, like high absorptions of water that increase the demand of water for specific workability.

Inadequate information on the durability and robustness perspective to the TRAC production is a thoughtful issue which needs complete attention because of its impacts on the changing aspects of the sector (Abdel-Shafy and Mansour 2018). It is currently unknown whether manufacturing, control of quality, and production costs in standard RCA production industries outweigh the advantages that originate from recycled concrete aggregates acquired as lower price constituent materials in concretes in comparison with natural concrete aggregates. One of the explanations behind the slow manufacturing acceptances for RCA production is the already existing massive ready-mixed concrete (RMC) production plants (Abdulmatin et al. 2019). Nevertheless, the present patterns in which TRAC is increasingly substituting NCA in multiple structural developments have gradually obtained importance because of specific reasons. TRAC production, for example, allows sustainability for concrete wastes and encourages recycling as opposed to disposals of landfills. Also, it has notably addressed scarcity of natural aggregates and minimized the natural aggregate demands, and ultimately, it has allowed the quarried natural aggregate conservation. Despite the costs, these and many other benefits have resulted in increasing concerns for the manufacturing and usage of TRAC in structural development. This research provides a detailed review of RCA use in material on the basis of the experimental results accessible in the printed studies. The most vital chemical, physical, and mechanical TRAC features are explained in this study. Nevertheless, more priority has been given to explain the TRAC effects on material durability and hardened and fresh characteristics (Al-Swaidani and Khwies 2018). The paper concentrated on the assessment of the ultra-performance concrete structures that favor the TRAC production activities rather than natural materials.

6.1.1 Definition and Significance of Tailor-Made Recycled Concrete Aggregates (TRACs)

Tailor-made recycled concrete aggregates (TRACs) originate from the Portland cement demolition. The pieces of aggregates can be anticipated to differ correspondingly given that the original concretes could have been weathered or fresh, open graded or dense, weak or strong. When the TRACs originate from central recycling plants, the consistencies shall have been tackled to certain level by materials' blending from diverse sources. The materials shall reflect more immediately and directly the concrete types being crushed if the materials are coming from on-site crushing plants. The crushing processes generate original concrete aggregate agglomerations with adhered mortars. Typically, these agglomerations are less conventional than angular aggregates. Form the mortar elements, the crushed concretes also shall generate fines.

Aggregates form the concrete skeleton. Often, aggregate occupies about seventy percent of the total concrete volume (Almeida and Cunha 2017). A greater percentage of the amount of these aggregates is mainly coarse aggregate. In the construction sector, the coarse aggregate demands are therefore enormous. The growing extractions of rough materials from the natural resources are needed to achieve these high demands. The ever-increasing applications of natural coarse aggregates (NCAs) create environmental imbalances. Therefore, the utilization of alternative rough materials is essential in the building industry. The application of TRAC acquired from the demolished concrete buildings is one of the approaches to accomplish this. The applications of reclaimed concrete materials in the constructions minimize the requirements for natural coarse aggregate. In turn, this minimizes the adverse ecological effects because of the extractions of natural aggregate. The growing charges for landfills and the NCA scarcity have also encouraged the TRAC use in concretes (Abdel-Hay 2017). Also, the increased distances between the building site and source of quality NCA have forced the contractors to consider replacing NCAs by TRAC.

The increasing ecological destruction concerns caused by the utilization of virgin aggregates have resulted in interest in more cost-efficient and conservative concrete materials of TRAC. TRAC has potentials to substitute NCA because of its low production costs and a wide variety of advantages. There is, nevertheless, minimal use of TRAC in the construction industry because contractors are still not aware of its cost-effective benefits and its procedures of production. Abdel-Hay argued that reinforced concretes with demolition wastes had been estimated to have almost fifty percent material wastes when quantified in terms of weights. Crushed concretes, signified as CC, are generated from the manufactured material wastes. The rough parts of it are removed and separated as they are viewed as coarse materials. The crushed part of the products is referred to as fine recycled concrete aggregates (FRCAs).

Nevertheless, structural grade concretes can also be generated by the applications of RCAs (Abdel-Hay 2017). The TRAC utilization in high-performance as well as high-strength structural concrete is possible through correct quality control and mix design and with the addition of silica fumes. Poverty has been noted to impose special restraints that result in delays and at times, no programs to execute different engineering processes (Bui et al. 2018). Cost issues limit communities from the implementation of environmentally and sufficient, friendly resolution like recycled concrete aggregate. The most vital chemical, physical, and mechanical TRAC features are explained in this study. Nevertheless, more priority has been given to illustrate the RCA effects on material durability and hardened and fresh characteristics. The significant findings of this study are that TRAC production setup can be utilized in large-scale production of recycled materials at low costs. Also, front-loader type and overhead bin kind of industries may be applied to reduce the TRAC incremental costs (Chatterjee and Sui 2019). Nevertheless, the pricing effects and demand and supply aspects of recycled materials pose different challenges that are rarely accounted for.

Various researchers have studied RCA features. Recycled concrete aggregates (RCAs) are an effective way to reduce constructions and demolition wastes (Abdel-Shafy and Mansour 2018). However, the features of concretes with RCA have to be explored. Especially, various elements such as replacement level, moisture content, type of TRAC, and water-to-cement ratios influence the TRAC concrete strengths. Different from natural aggregate, TRAC has typically higher water absorption, which could significantly affect concrete properties, especially the workability. Further, concrete with drier RCA had a more considerable slump and faster slump loss than that of saturated TRAC. It is clear that the strengths for the entire fine aggregate replacement increase with time with the blends having higher percentages of the combination gaining at a higher percentage. This depicted the characteristics of the presence of lateritic material which from the literature review available, it stated that the strength for the material increase with the structure's life.

Currently, concrete is the most basic construction material (Abdulmatin et al. 2019). The production of concrete uses up a lot of resources mainly sand where its harvesting has a great environmental impact on riverbeds and sand borrow pits, hence the need to innovate methods to supplement concrete production with other substances like TRAC from demolished sites and heap of waste mortar in mega projects blended with laterite soil as a sand substitute. In the recent past, concrete structures have been undergoing demolition resulting in the debris which commonly has been utilized as landfills (Ahrari and Haghani 2019). The environmental sustainability is a key focus, and the need to keep construction costs low, the debris can be crushed to achieve the fine recycled concrete aggregates (FRCAs) which can be utilized in replacing the fine natural aggregates (sand) partially. This can be used to produce fresh concrete, therefore achieving a sustainable construction industry. In the recent past, concrete structures have been undergoing demolition resulting in the debris which commonly has been utilized as landfills (Abdel-Hay 2017) partially. This can be used to produce fresh concrete, therefore achieving a sustainable construction sector.

As the aggregate may be retrieved from concrete residuals after placing is finished, waste generation need not to be a challenge within the construction sector (Abdel-Shafy and Mansour 2018). Usually, this represents the concrete sludge and 2.00% of the aggregate fresh material remaining after the reclamation of the aggregates. The remaining concrete sludge can be mixed with the liquid from washing of the truck. The slurry of cement pastes can be dehydrated, desiccated, grounded, and utilized to substitute NCAs with fine TRAC. The only sure way of minimizing construction costs is the utilization of the available local resources and adoption of innovative construction materials (Al-Bayati et al. 2016). Applying TRAC and blended laterites as fine aggregates on all these accounts in concretes appear to the best, particularly in the place in which the mixed substances are freely available other than the natural concrete aggregates such as river sand.

6.1.2 Characteristics of Ultra-Performance Concrete Made with TRAC

6.1.2.1 Physical Characteristics

The concrete properties and mix proportions are influenced by the physical TRAC characteristics. The fundamental properties like absorptions, pore volumes, bulk density, specific gravity, textures, and shapes of TRACs are overall inferior than those of virgin concrete aggregate because of the impurities and residual cement mortar/paste presence (Andal et al. 2016). The magnitudes of the impacts differ with quantities and nature of recycled cement mortars or pastes that are present in TRAC. From the available literature, the summary of fundamental physical TRAC features is indicated in Table 6.1 and thereafter explained.

(1) **Particle shapes and textures**

The TRACs appear to be very course and angular because of the grinding of the concretes and due to the presence of toughened cement mortars or pastes adhered to original coarse aggregate surfaces (Andal et al. 2016). Depending on the sizes of aggregates, particles of TRACs typically have thirty to sixty percent mortars or pastes. Higher amounts of old cement pastes or mortarsattached small size proportions of rough aggregates. In particle shape, recycled concrete aggregates are identical to grinded rocks, but the types of grinding tools influence the gradations and other features of fine concretes.

Table 6.1 Fundamental physical characteristics of NCAs and TRAC (Andal et al. 2016)

Physical characteristics	Natural concrete aggregates	Tailor-made recycled concrete aggregates (TRACs)
Pore volumes (volume percentage)	0.50–2.00	5.00–16.50
Absorptions (wt%)	0.50–4.00	3.00–12.00
Compacted bulk densities (kg/m^3)	1450.00–1750.00	1200.00–1425.00
Specific gravity (soaked surface-dry-oriented)	2.40–2.90	2.10–2.50
Shapes and textures	Smooth and well-rounded sand to course and angular grounded rocks	Angular with course surfaces

(2) **Specific gravity**

The TRAC specific gravity is often lesser that that of virgin concrete aggregates. The lesser TRAC specific gravity is because of old cement mortar/paste's presence of the particles of aggregate which makes it less dense than NCAs due to higher porosities (Arredondo-Rea et al. 2019). In saturated surface-dry conditions, the distinctive TRAC specific gravity values vary from 2.10 to 2.50 as shown in Table 6.1. These are 5.00–10.00% lesser that that of NCAs.

(3) **Bulk density**

As illustrated in table one, the TRAC bulk densities are greatly lesser than that of virgin concrete aggregates. The TRAC bulk densities according to the experimental findings of Aliabdo et al. (2018) have been established to be 9.80% lesser than that of virgin gravel aggregates. This is majorly because of the higher TRAC porosities in the adhered cement mortars/pastes' presence.

(4) **Pore volume**

The TRAC pore volumes are significantly greater than that of NCAs as shown in table one (Andal et al. 2016; Arredondo-Rea et al. 2019). This is because of the adhered cement pastes/mortars' porous nature emerging from old concretes. As compared to virgin concrete aggregates, the TRAC higher pore volumes make it less dense and weak.

(5) **Absorption**

As can be seen in table one, the TRAC absorptions are substantially greater than that of virgin concrete aggregates (Andal et al. 2016; Aliabdo et al. 2018; Arredondo-Rea et al. 2019). A specific cement mortar/paste quantity from the initial concrete remains fixed to the particles of TRACs when demolished concretes are grounded. As compared to the initial aggregates, the attached cement mortars/pastes possess higher porosities. This is the major reason for the greater TRAC absorptions.

6.1.2.2 Mechanical Characteristics

The concrete mechanical features rely on the aggregate mechanical characteristics. It was established that TRAC mechanical features are lower as compared to the mechanical features of virgin concrete aggregates from the available literature (Andal et al. 2016; Aliabdo et al. 2018; Arredondo-Rea et al. 2019). The major TRAC and NCA mechanical characteristics are indicated in Table 6.2 and thereafter explained briefly. The TRAC concrete strengths may be boosted by either allowing reclaimed aggregates to soak up parts of mixing water without or with pozzolanic liquids during mixings or soaking the TRACs in mixes of pozzolanic liquids such as colloidal silicas or water before mixing of concretes. The micro-cracks in TRACs are expected to be filled up by the absorbed pozzolanic liquids or absorbed water with cement gels during pozzolanic reactions or cement hydrations. Therefore, the TRAC concrete strengths can be increased.

(1) Aggregate abrasion values

The aggregate abrasion values (AVVs) are measures for aggregate wear resistance. When the material loss because of wear becomes higher, a greater AVV is obtained. Usually, the TRAC aggregate abrasion value is greater than that of NCAs. The classic Los Angeles TRAC value varies from 20.0 to 45.0% that is greater than the values of virgin concrete aggregates as indicated in Table 6.2 (Bui et al. 2018; Dimitriou et al. 2018; Hou et al. 2019; Mohammed and Najim 2020). Despite of their origins, the TRAC aggregate abrasion values nevertheless are basically below the acceptable optimum limits for structural applications (fifty percent by weights).

(2) Aggregate effect values

The aggregate effect values (AEVs) are aggregate strength values exposed to impacts. AEVs indicate the aggregate resistances to dynamic loads. As shown in table two, the NCA aggregate impact values (15–20%) are lower than that of RCA (20–25%) (Bui et al. 2018; Dimitriou et al. 2018; Hou et al. 2019; Mohammed and Najim 2020). The attached cement and mortar pastes make TRACs less strong and thus lead to a higher aggregate impact values for recycled concrete aggregates.

Table 6.2 Main TRAC and NCA mechanical characteristics (Arredondo-Rea et al. 2019)

Mechanical properties (wt.%)	Tailor-made recycled concrete aggregates	Natural concrete aggregates
Aggregate grounding values	20–30	14–22
Aggregate abrasion values	20–25	15–30
Aggregate effect values	20–25	15–20

(3) **Aggregate grounded values**

The aggregate grounded values (AGVs) offer measures for aggregate resistances to grinding under progressively applied compressive loads. The aggregates are stronger when the value is lower (Bui et al. 2018; Dimitriou et al. 2018; Hou et al. 2019; Mohammed and Najim 2020). It has been established that ACVs of NCA (14–22%) are substantially lesser than that of TRAC (20–30%) from the literature and as indicated in table two. This is anticipated due to the comparatively weak mortars and cement pastes fixed to the particles of TRACs.

6.1.2.3 Chemical Properties

The RCA chemical characteristics influence the RCA concrete performance. The concrete durability shall be affected if the reactive chemicals such as chlorides, alkalis and sulfates are present in recycled concrete aggregates. The RCA chemical features are briefly explained below.

(1) **Soundness**

Aggregate soundness is indications of resistances of aggregates to environmental effects such as weathering (Poon and Chan 2007; Tabsh and Abdelfatah 2009); Anderson et al. 2009; Andal et al. 2016). The soundness test of magnesium sulfates and sodium sulfates is basically applied in assessing the aggregate soundness. TRACs commonly pass soundness tests of magnesium sulfates, but fail soundness tests of sodium sulfates. For TRACs and NCAs, respectively, the soundness losses of magnesium sulfate obtained by Anderson et al. (2009) were 3.70 and 2.50%. These figures are identical to the findings acquired by Tabsh and Abdelfatah (2009). Poon and Chan examined soundness tests of sodium sulfates and established the losses were 12.0 and 9.0% respectively for TRACs and NCAs. Therefore, sodium sulfate soundness test met the standard requirements. Also, the findings are identical to the results reported by Andal et al. (2016). The TRAC unsoundness is majorly associated with the distributions of the pore size in the aggregates (Andal et al. 2016).

(2) **Reactivity of Alkali–aggregates**

In concretes, TRACs can cause alkali–aggregate reactions (AARs) if the original concrete aggregates were prone to alkali–aggregate reactions. Besides, the mortars adhered to RCAs or cement paste alkali content might have substantial effects on new concretes' AAR susceptibility having TRACs. The TRAC alkali contents are linked with contents of its mortars. Recycled concrete aggregates having higher quantities of mortars lead to greater contents of alkali, and thus they are more susceptible to alkali–aggregate reactions (AARs).

(3) **Sulfate contents**

RCAs can have great sulfate contents because of the sulfate substances present in the adhered mortar cements. Andal et al. (2016) investigated the contents of water-soluble sulfates for RCAs and NCAs. The researchers established sulfate contents of 0.032 g/L and 0.0250 g/L for RCAs and NCAs, respectively. The RCA sulfate contents depend on the adhered mortar/cement paste amounts. Higher RCA sulfate contents reveal that higher mortar amounts are attached to RCAs.

(4) **Chloride contents**

High degrees of chlorides have been established in the RCAs generated from the bases with lasting exposures to chloride-oriented unfreezing substances. The RCAs with higher contents of chlorides can influence the new concrete durability because of the steel reinforcement corrosions (Poon and Chan 2007; Anderson et al. 2009; Tabsh and Abdelfatah 2009; Andal et al. 2016). Because the increased corrosions of steel might result in the early reinforced concrete structure failure, the RCAs obtained from the old concretes having chloride higher than 0.04 kg/m^3 should not be applied in new concretes.

(5) **Organic matters**

Polymeric materials, rubbers, plastics, joint seals, textile fabrics, woods, and paper can be present in recycled concrete aggregates (Poon and Chan 2007; Anderson et al. 2009; Tabsh and Abdelfatah 2009; Andal et al. 2016). In concretes, these materials become unbalanced when exposed thawing, drying, and freezing or wetting. 0.150% by RCA weight is the acceptable organic matter limit.

6.2 Deterioration Mechanisms and Micro-nanostructure-Related

There are two methods for fabrications of nanostructures and syntheses of nanosubstances. Top-down techniques strive to establish by utilizing bigger one direct their assemblies. Many approaches that originated from traditional solid-state silicon technologies for fabrication of microprocessors are now able to create properties lesser than 100 nm, falling under the nanotechnology definitions. As atomic layer deposition (ALD) approaches, massive magnetoresistance-oriented hard drive now in the markets fits these descriptions. Also, solid-state methods can be employed in creating NEMS (nanoelectromechanical system) that are associated with MEMS or micro-electromechanical system. At the same time, concentrated ion beams may deposit materials or remove materials when appropriate pioneer gases are used. These techniques for instance are applied regularly to establish sub-100-nm material sections for transmission electron microscopy analyses. To deposit a resist, atomic force microscopes may be utilized as nanoscale write heads that are then accompanied with etching processes to eliminate materials in top-down techniques.

Sand being one of the ingredients in the production of concrete is sourced from river beds and borrows pits, which poses a great danger to the environment and huge cost incurred during transportation from the sources (Bartolacci et al. 2018). The modifications in the establishments of RMC plants are examined, and the findings are noted to favor RCA production that relies on aggregate feedings and storage mechanism of RMC plants. The RMC industries in this research are then categorized into two diverse types based on the results, namely front-end loaders (FELs) and overhead bins (OBs) plants (Ding et al. 2016). The overhead bin facilities apply aggregate bin for storing that are suited either above ground, on the grounds or undergrounds to collect materials, store them, and utilize them in programmed processes that weigh and feed in different lots. Front-end loaders receive the aggregate and store it in ground-based compartment that is open. The front-end loaders in this process manually feed the aggregate in batch directed to weigh hoppers in processes that are ultimately programmed. Because of the process simplicity, the present facilities may be categorized into the first two plants. Reviews of overhead bin plants suggest that they are majorly used in municipalities because of their high capacities of production (Du and Lin 2018). On the other hand, front-end loaders are applied in peripheral and regional places because of their comparatively low capacities of production.

Various blends with water and superplasticizer variations are utilized as trial judges of rendering the testing robust. Various rheological features, workability properties, and mechanical characteristics are the parameters to be tested. The strength of different blends having variations of cements is tried for robustness (Bartolacci et al. 2018). The applications of high-level recycled aggregate have been determined to adversely influence the concrete characteristics. The elastic static and strength modulus reduce with the increases in the recycled material contents.

The applications of the high performance of recycled materials have been determined to adversely influence the concrete features (Wijayasundara et al. 2017). The elastic static modulus and the compressive strengths reduce with the increases in the recycled material contents. Nevertheless, the reductions can be sufficiently compensated by the applications of low w/b ratio. Besides, cement additions in the recycled aggregates at the same replacements and w/b ratio reduce the tensile splitting strengths and the compressive strengths. Nevertheless, partial fly ash applications as cement replacements decrease the RCA shrinkage (Wijayasundara et al. 2016). The fly ashes result in the long-term and greater RCA durability because of pozzolanic fly ash reactions.

6.3 Impermeability and Water Absorption

6.3.1 Permeability

Concerning the RCA impacts on the concrete porosity, restricted research has been found. Chatterjee and Sui (2019) established grade twenty to forty concretes and

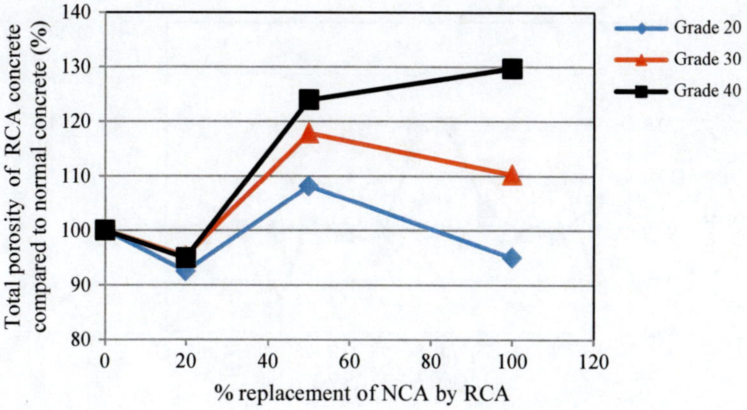

Fig. 6.1 RCA effects on the aggregate concrete porosity (Deakins and Bensemann 2019)

illustrated that at RCA 20% substitutions. As indicated in figure one, the aggregate porosities were lesser than that of the control concretes with virgin concrete aggregates. Also, the researchers reported that, for all concretes, the aggregate porosities were increased at content of 50% RCAs. At 100% of NCA substitution, grade 20 RCA concretes however had marginally lesser porosities than the concretes of 20 NCAs. The RCA concrete porosity can be 10–30% greater than those of NCA concretes depending on the grade of strength according to Deakins and Bensemann (2019) as shown in Fig. 6.1. Such RCA concrete performances are because of the differences in the compositions of the mixes of concrete such as RCA and cement content, aggregate quantity, and pozzolanic percentages.

The concrete permeability relies on both the concrete matrix permeability (binder and cement pastes) and the included aggregate absorption capacities. Also, the concrete permeability is influenced by its continuities of the pores, distributions, and size and porosity. For equivalent water-to-cement ratios, the RCA concrete permeability seems to be greater than those of NCA concretes. The RCA concrete permeability may be 200–500% greater than those of concretes of NCAs. As indicated in Fig. 6.2, the mixes of RCA concretes if correctly accounted for in designs may be designed to have an identical permeability as those of NCA concretes.

6.3.2 *Water Absorptions*

To establish the RCA concrete absorptions, restricted studies have been conducted. The RCA concrete water absorption generally is anticipated to the greater than those of concretes of NCAs. This is because of the substantially high air permeability and water absorptions of RCAs. As indicated in Fig. 6.3, Forsyth (2018) created different concrete grades and indicated that the concrete water absorptions increased with the

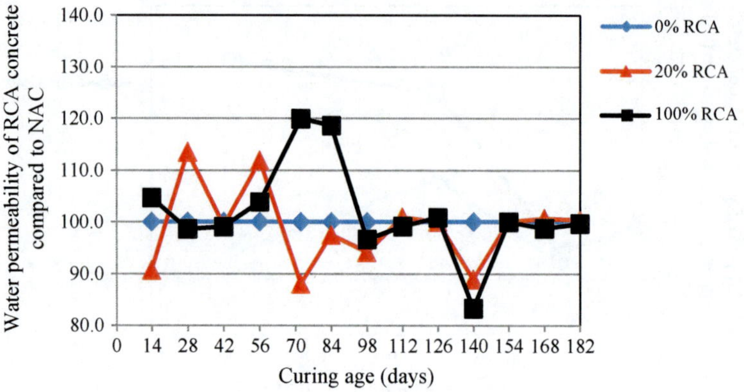

Fig. 6.2 RCA effects on the concrete water permeability (Deakins and Bensemann 2019)

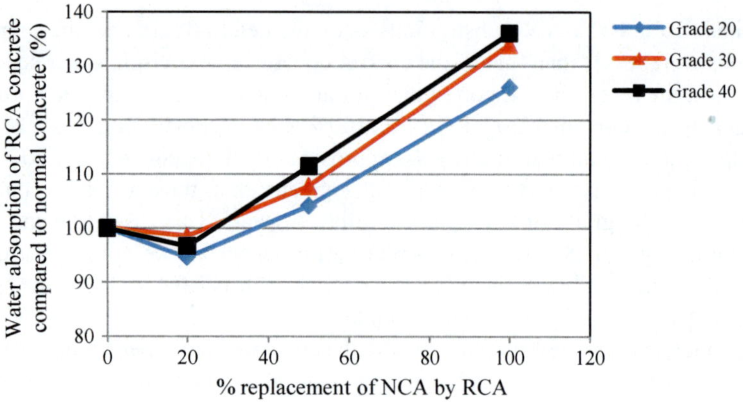

Fig. 6.3 RCA effects on the concrete water absorptions (Forsyth 2018)

increase in content of RCAs except for 20% substitutions. The RCA concrete water absorptions are 0–40% greater than those of NCA concretes relying on the strength grades according to Forsyth (2018). The author did not explain the basis for lesser absorptions of water at the 20% RCA contents, but suggested further investigations.

6.4 Durability Properties

6.4.1 Sulfate Resistance

Limited studies indicated that the RCA concrete resistances to sulfates are about similar or marginally lesser to those of natural concrete aggregates. Generally, RCA concrete sulfate resistance may be enhanced by utilizing silica fumes, fly ashes, and fine granulated blast furnace slags and with correct proportion of mixtures. The embedded steel corrosion rates may be decreased in RCA concretes by good construction quality controls and correct proportioning of mixtures. Meneses et al. (2018) applied polarization approaches in identifying the corrosion behaviors of steel fixed in RCA concretes. The researchers reported that rates of corrosions never seemed to rely on aggregates for lower chloride levels (0.50%) and the binder type. As compared to NCA concretes, the investigators also reported that RCA concretes with 65.0% fine granulated blast furnace slags and 30.0% crushed fuel ashes were more successful in mitigation of corrosion reactions at greater levels of chlorides.

6.4.1.1 Chloride Resistance

Few studies have been carried out on RCA concrete chloride penetration resistance. Hou et al. (2019) noted that high-performance concrete chloride penetration values made without and with RCAs were very identical. The 100% RCA applications in concretes decrease their capacities to resist chloride ion penetrations by about 30.0% in comparison with concretes of NCAs according to Jin et al. (2017). Nevertheless, the RCA concrete chloride penetration resistances can be enhanced greatly by increasing extra cementing substances like ground granulated blast furnace slags and fly ashes. Júnior et al. (2019) disclosed that as compared fly ashes, ground granulate blast furnace slags to improve the RCA concrete chloride ion penetration resistances. As compared to the fly ashes, the slag decreased the concrete chloride diffusion coefficients by around 120–200% in their research.

6.4.2 Carbonation Resistances

On the concrete carbonation, the existing studies indicate the contradictory RCA effects. Mohammed and Najim (2020) reported that depths of carbonation decrease when contents of RCAs increase, giving better performances except for replacements of 100%. RCA concretes need higher contents of cements to accomplish the identical NCA concrete strengths at the same water-to-cement ratios according to Mohammed and Najim (2020). The higher cement contents have greater reserves of alkalis that act to carbonation depth in concretes. O'Donnell et al. (2019) in contrast found that at the same water-to-cement ratio, RCA concrete carbonation depth is greater than those of NCA concretes. Besides, the RCA concrete carbonation rates are four times higher than those of NCA concretes. The greater carbonation can increase the reinforcing steel corrosion risks in RCA concretes. Such risks may be counterbalanced by increased concrete covers, appropriate extra cementing materials, correct curing, and lower water-to-cement ratios. More studied are however required to analyze the RCA effects on the concrete carbonation resistance.

To examine the RCA concrete performance against alkali–silica reactions, insufficient studies have been carried out. RCA concrete resistances to alkali–silica reactions depend on the RCA sources. Poon et al. (2003) noted that RCAs obtained from ancient concretes having alkalis affect the newly created RCA concrete resistances to alkali–silica reactions. The findings of the researchers showed that the new concretes prepared with RCAs increased disproportionately because of alkali–silica reactions. The low-lime Class-F fly ash applications may significantly minimize the expansions because of alkali–silica reactions in RCA concretes. The existing information is inadequate to reach specific conclusions on the RCA concrete alkali–silica reactivity.

On the RCA concrete durability under acid attack, few studies are available. Poon and Chan (2007) noted that the RCA concrete resistance-to-acid attacks are similar to or marginally lower to those of NCA concretes. The scholars immersed the concrete specimens on testing baths having sulfuric acid at a pH of 2 to examine the RCA concrete degradation level against acid attacks. As compared to NCA concretes, the penetration of acids into RCA concretes was marginally greater. Nevertheless, the investigators illustrated that RCA concrete resistance-to-acid attacks can be enhanced by additions of fly ashes. Additional RCA concrete studies subjected to acidic conditions are needed to examine their durability properties because of few studies.

6.4.3 Water Attack

Generally, the NCA concrete bleeding is greater than that of RCA concretes (Abdel-Hay 2017). During mixing, some of the ancient cement pastes rubbed off recycled concrete aggregates and generate extra fines in the mixes of concretes. These fines therefore minimize the concrete bleeding after adsorbing some of the mix water. At lower free water contents, the higher quantities of fines also increase the concrete mix cohesiveness. Besides, the increased surface coarseness and angularity at greater RCA contents contribute to the improvements of concrete cohesiveness. The concrete mix stability enhances because of improved cohesiveness and minimized bleeding (Abdel-Shafy and Mansour 2018). Therefore, the RCA concrete segregation resistance can be compared to that of virgin concrete aggregates as shown in table three.

Various investigators noted that RCA concretes show enough resistances to cycles of freeze–thaws. There are proofs that RCA concrete multiple recycling further improves frost resistances. For RCA concretes, some scholars reported about similar or marginally reduced frost resistances as compared to concretes of NCAs. Besides, concretes made with dry and saturated RCAs portrayed weaker resistances of freeze–thaws. The RCA concretes however prepared with semi-saturated concretes indicated best outcomes because of the enhanced attachments at the interfaces between pastes and aggregates. However, more studies are required to confirm the RCA impacts on the concretes' freeze–thaw resistances.

The concrete capacities to resist abrasions, chemical attacks, weathering actions, and other adverse conditions of services are durability. Even when the RCA is generated from concretes with durability problems, RCA concretes can be significantly durable, provided that the food quality is sustained during construction and proportion of mixtures is carried out correctly (Poon and Chan 2007; Anderson et al. 2009; Tabsh and Abdelfatah 2009; Andal et al. 2016). The diverse RCA concrete durability characteristics are explained below.

Primarily, drying shrinkages are functions of water-to-cement ratios and paste contents and controlled by particles of aggregates. Some works have found 20% to 50% greater shrinkage than NCA concretes because RCA concretes possess greater paste contents (Poon and Chan 2007; Anderson et al. 2009; Tabsh and Abdelfatah 2009; Andal et al. 2016). Few works in contrast noted relatively lesser values of RCA concrete drying shrinkages. At different concrete curing ages, Poon and Chan (2007) reported low shrinkage strains that was made substituting 30% virgin concrete aggregates by recycled concrete aggregates as illustrated in Fig. 6.4. The contradicting findings propose that more studies are needed for investigating the RCA effects on concrete dying shrinkages.

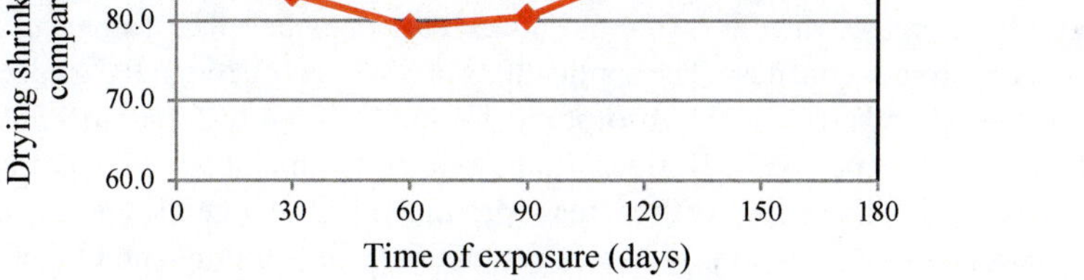

Fig. 6.4 RCA effects on concrete drying shrinkages (Poon and Chan 2007)

6.5 Service Life Prediction of Ultra-Performance Concrete (UPC) Containing TRAC

The RCA concrete features are affected by various main aspects like air entrainments, cement contents, curing conditions, the RCA moisture conditions, the qualities of the parent RCA concretes, the RCA physical features, the RCA size and type, the content of RCA, and the water-to-cement ratios (Poon et al. 2002). The aspects are explained below.

6.5.1 RCA Features

The RCA physical characteristics substantially affect the hardened and fresh concrete properties. For example, the mixes of fresh concretes with coarse-textured and angular particles become harsh and thus are hard to complete (Poon et al. 2004). Also, the concrete finishability and workability can be influenced because of the high RCA absorption. Besides, the greater RCA pore volumes can influence transportation features (permeability and water absorption), strengths, and the porosities of the concretes.

6.5.2 The Contents of Recycled Concrete Aggregates (RCAs)

The toughened and fresh features of concretes are highly affected by the contents of RCAs utilized as full or partial NCA replacements. Using seven independent variables, Poon et al. (2002) created a model regarding the amount and type of aggregate to predict the performances of concretes for 0.0–100% NCA replacements with RCAs. The researchers reported that the concretes generated with RCAs had lesser elastic modulus and compressive strengths than NCA concretes. The greater RCA contents also increase water absorptions but decrease densities, therefore resulting in increased concrete porousness. The applications of 50%–100% rough RCAs decrease concrete density and increase the absorption of water by roughly 2.11% to 3.50% and 0.14% to 0.38%, respectively. Besides, Ratnayake and Samarakoon (Ratnayake and Samarakoon 2017) noted that with increase grounded RCA contents, the resistances to penetrations of chloride ions, whereas concrete tensile splitting and compressive strengths reduced. Also, the investigators that concrete drying shrinkages improved with increases in RCA contents. It can cover be regulated by lowering water to cement ratios.

6.5.3 Parent Concrete Quality

Rattanachu et al. (2019) explore the influences of quality of parent concretes on the RCA concrete features. The scholars reported that the RCA water absorptions increase with the improved parent concrete strength. This is because the concretes with greater strengths fundamentally need higher cement contents, therefore increasing the mortar amount sticking to the aggregate. The mix water content adjustments therefore are necessary for the ne concretes, including the RCAs obtained from old concretes of greater strengths to derive the preferred workability. The porous RCAs affect the RCA concrete strengths. The proportional loss in the tensile or compressive strengths of new concretes because of the RCA use is more substantial when it is obtained from weaker old concretes than strong old concretes (Tabsh and Abdelfatan 2009).

6.5.4 Sources or Types of Recycled Concrete Aggregates (RCAs)

Few works have been conducted to explore the effects of RCA sources or types on characteristics of concretes. Except for the concretes prepared of recycled grounded aggregates obtained from brick ceramics in which increases in compressive strengths were reported, the concretes generated with rough RCAs had a lesser compressive strength according to Teh et al. (2018). The concrete elastic modulus was decreased for all RCA types. The RCAs derived from red ceramics, however, exerted greater influences in decreasing elastic modulus of concretes because of lower densities.

6.5.5 Sizes of Recycled Concrete Aggregates (RCAs)

Thongkamsuk et al. (2017) applied three diverse sizes of aggregates in assessing the RCA size influences on the concrete properties. The greater reductions in the elasticity modulus were derived for the concretes prepared with tinier RCA sizes. On the other hand, they reported that the strengths increase with increases in the optimum RCA sizes. Also, they found that the concrete water absorptions reduce with the increases in the maximum RCA sizes. This is because of the comparatively lower contents of weak mortars stuck to large-sized aggregate.

6.5.6 RCA Moisture Conditions

The aggregate moisture conditions affect the concrete workability. The concrete original slumps (workability measurements) are greatly reliant on the original free water contents of the concrete mixes. Verian et al. (2018) revealed that while the air-dry and saturated surface-dry RCA shows ordinary initial slumps and slump losses, the oven-dry RCAs result in faster slump losses and greater original slumps.

6.5.7 Curing Conditions

More negative effects on RCA concretes are generated by the outer environmental curing condition than on NCA concretes. Verian et al. (2018) revealed that the differences in the splitting tensile strengths among the concretes of NCAs and RCAs are high when they cured in the external environments. Besides, Wang et al. (2016) revealed that the water-cured RCA concrete carbonation depth is virtually double than those of concretes of air-cured RCAs. The reductions in the carbonation depth generated by water-curing could be partly because of greater concrete internal humidity.

6.5.8 Cement Contents

With high cement contents, the RCA concretes were reported to have high carbonation resistances. Wang et al. (2016) suggested greater RCA concrete cement contents to accomplish preferred compressive strengths. The RCA concrete tensile resistances increase with improved cement contents in concretes according to Wijayasundara et al. (2017).

6.5.9 Water-to-Cement Ratios

The decrease in the RCA concrete performances is associated with the water-to-cement ratios utilized in mix designs. Wijayasundara et al. (2018) reported that as compared to the RCA parent concretes, RCA concretes need greater cement contents and lesser water-to-cement ratios to accomplish specific compressive strengths. The NCA resistance to thawing and freezing at water-to-cement ratios of 0.290 was exceptionally high. But for RCA concrete, the same water-to-cement ratios failed to offer suitable freezing–thawing resistances.

6.5.10 Air Entrainments

Through appropriate air entrainments, durable RCA concretes can be generated. The air entrainments are as successful for NCA concretes as for RCA concretes. Besides, entrained air application is more successful than reducing the water-to-cement ratios to enhance RCA concrete resistances to thawing and freezing.

6.6 Physical Properties of UPC Containing TRAC

6.6.1 Fresh Features

The RCAs can affect the fresh concrete characteristics because of their higher porosities, absorptions, surface coarseness, and angularity (Poon and Chan 2007; Anderson et al. 2009; Tabsh and Abdelfatah 2009; Andal et al. 2016). As realized from the available literature, the RCA effects on the major concrete fresh features are provided in table three and explained thereafter.

6.6.2 Workability

The higher coarseness and angularity particles of recycled concrete aggregates reduce the concrete workability and make them more hard to complete appropriately (Poon and Chan 2007; Anderson et al. 2009; Tabsh and Abdelfatah 2009; Andal et al. 2016). The reduction levels in workability increase with the improved RCA proportion in the mixes of the concretes. More water is thus needed for RCA concretes to get similar NCA concrete workability. As can be noticed from table three, the mixes of concretes integrating RCAs generally fulfill the original requirements of slumps. Nevertheless, the greater RCA absorptions can result in a rapid workability loss, therefore restricting the period needed for putting and completing of concretes (Abdulmatin et al. 2019). The challenges related to the rapid workability loss should be tackled by modifying and managing the RCA water content before mixings, not by addition of more water at the worksites.

6.6.3 Wet Density

To examine the RCA effects on the concrete wet density, several studies were conducted. Generally, the RCA concrete wet density is lesser that of virgin concrete aggregates as noticed in table three (Ahrari and Haghani 2019). As compared to NCA concretes, it has been noticed that RCA concrete wet density is 5–15% lesser. This is RCAs possess adhered old cement pastes or mortars whose densities are lesser than the NCA densities.

6.6.4 Air Contents

The air contents of concretes are substantially affected by the volumes of their mortars. The RCAs influence the air contents of concretes since they have bigger mortar contents. As can be noticed in table three, the fresh concrete air concretes having RCAs are usually up to 0.60% greater than the air contents of the fresh NCA concretes and slightly more variable (Al-Bayati et al. 2016). The air that entrapped and entrained in the recycled RCA mortars causes this. While deciding the target RCA concrete air content, the existing mortar air content therefore.

The RCA effects on the toughened concrete characteristics may be substantial or insignificant relying on their physical features, gradations, contents, types, and sources. The toughened RCA concrete characteristics decrease with the NCA substation level by reclaimed concrete aggregates (Al-Bayati et al. 2018). Without greatly affecting the hardened concrete properties, up to 30% of virgin concrete aggregates by RCAs on weight basis as a general principle. As noted from the available literature, the variety of transformations in the toughened concrete features because of RCAs is provided in Table 6.3. The major RCA concrete hardened characteristics are explained below.

Table 6.3 RCA effects on the concrete hardened features (Al-Bayati et al. 2018)

Properties	Ranges of changes
Thermal expansions	10–30% more
Creeps	30–60% more
Drying shrinkages	20–50% more
Penetrations of chlorides	0–30% more
Water absorptions	0–40% more
Permeability	0–50% more
Porosities	10–30% more
Elasticity modulus	10–45% more
Strength of bonds	9–19% less
Flexural strengths	0–10% less
Splitting tensile strengths	0–10% less
Compressive strengths	0–30% less
Dry densities	5–15% less

6.6.5 Dry Densities

Typically, the RCA concrete dry densities are 5.0–15.0% less than that of virgin concrete aggregates. This is due to the recycled mortars attached to the RCAs. Depending on the aggregate size, the mortar quantities attached to recycled concrete aggregates differ in the ranges of 30–60% by RCA volume (Al-Swaidani and Khwies 2018). The recycled mortar density is much lesser than those most natural concrete materials. This leads to a lesser RCA concrete density.

6.6.6 Compressive Strengths

As indicated in Fig. 6.5, the compressive RCA concrete strengths are often lesser than those of virgin aggregates. The compressive RCA concrete strengths most commonly are 5–10% lesser than those of natural concrete aggregates. Depending on the RCA quality, it may also be reduced up to 25%. Normally, the greater air contents found in the mixes of concretes having RCAs can also result in lesser values of strength (Aliabdo et al. 2018). RCA concretes nevertheless can have the same and at times greater compressive strengths than virgin aggregates if the recycled concretes are obtained from old concrete sources that were initially generated with lesser water-to-cement ratios than the new concretes.

Anderson et al. (2009) established that by weights, RCAs generated no effects on the concrete compressive strengths up to the substitution degree of 30%. For the RCA content more than 30%, the compressive strengths reduced. Andal et al. (2016) revealed that the concrete compressive strengths were much lesser when RCAs were utilized in the oven-dry states. 20–30% reductions in compressive strengths were realized because of the RCA applications in the case of high-performance concretes. Similar results were noted by other investigators (Andal et al. 2016; Aslani et al.

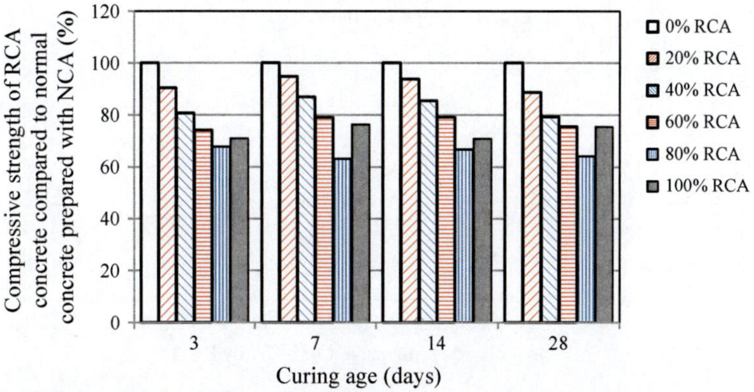

Fig. 6.5 RCA effects on the concrete compressive strength (Balan and Thomas 2019)

2018; Arredondo-Rea et al. 2019). Balan and Thomas (2019) explored the self-compacting features of concretes produced with RCAs and compared with properties of NCA concretes. The investigators established that at the same age, the variation in compressive strengths was insignificant.

Also, the fine RCAs can influence the concrete compressive strengths. The RCA concrete compressive strength is affected by the rough concrete-to-fine concrete ratios of the RCA source concretes according to Bui et al. (2018). The lesser coarse-to-fine aggregate ratios result in greater mortar quantities attached to particles of coarse RCA and therefore lead to reductions in the RCA concrete strengths. These decrements are even higher when reclaimed fines are utilized. Thus, the fine RCA uses in concretes are generally not recommended. Nevertheless, no reduction in compressive strengths took place for the concretes having up to 20% fine RCAs according to Di Maria et al. (2018). The strengths reduced with increases in the contents of RCAs beyond this level.

6.6.7 *Splitting the Strengths of Tensile*

Regarding the RCA effects on the concrete splitting tensile strengths, restricted literate is available. As indicated in Fig. 6.6. Dimitriou et al. (2018) found that RCA concrete splitting tensile strengths are lesser than those of NCA concretes. Various investigators reported that the RCA concrete splitting tensile strengths are 0–10% lesser than those of NCA concretes. During the time of 90–365 days, no statistically notable reductions in tensile strengths took place. Guo et al. (2018) on contrary noticed that RCA concretes generated greater tensile strengths than concretes of NCAs. Therefore, more studies are needed to examine the RCA impacts on the concrete splitting ductile strengths.

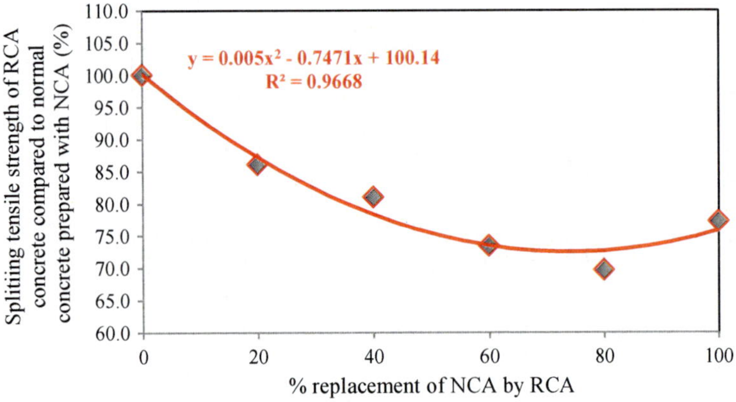

Fig. 6.6 RCA effects on the concrete splitting ductile strengths (Guo et al. 2018)

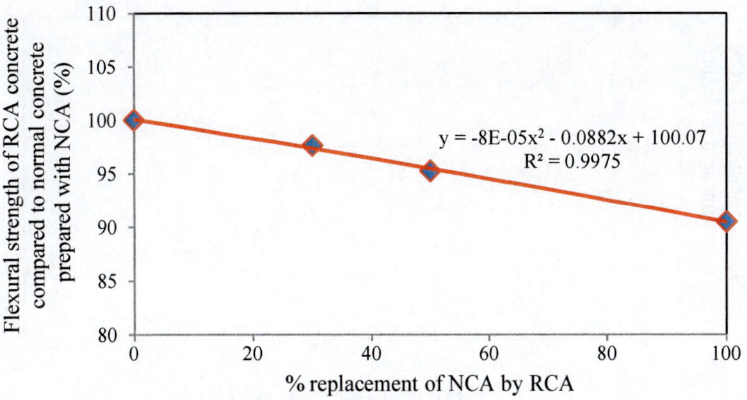

Fig. 6.7 RCA effects on the concrete flexural strengths (Durdyev et al. 2017)

6.6.8 Flexural Strengths

Generally, as indicated in Fig. 6.7, the RCA concrete flexural strengths are lesser than those of NCA strengths (Fraile-Garcia et al. 2017). Typically, the RCA concrete flexural strengths are lesser than those of NCA concretes. The 3-day RCA concrete flexural strengths were greater than those concretes of NCAs according to Ding et al. (2016). At the period of twenty-eight days, the strength was lower. In their research, the NCA concretes increased strengths progressively and had greater flexural strengths than the concretes of RCA at later ages. Durdyev et al. (2017) noted that RCAs never produced any notable adverse effects on the concrete flexural strengths. However, with adequate strengths, RCA concreters can be generated for various uses, at times even with 100% substitutions of NCAs.

6.6.9 Bond Strengths

The concrete bond strengths are indications of the interconnecting features of pastes and aggregates. The coarse RCA surfaces lead to better bond than virgin concrete aggregates. For testing the bonds between reinforcements and concretes that included 100%, 50%, and 0% RCAs, Du and Lin (2018) used 150 mm × Ø100 cylindrical samples with fixed mild and ribbed reinforcements (diameter = 12.00 mm and embedded lengths = 150 mm). The findings indicated that the bonds between reinforcements and RCA concretes are not greatly affected by the RCA inclusions in concretes. The findings from Babel et al. (2020) nevertheless disclosed that the NCA concrete bond strengths were 9.0–19.0% greater than those of RCA concretes. Such contradicting findings mean that more studies are needed to examine the RCA effects on the concrete bond strength.

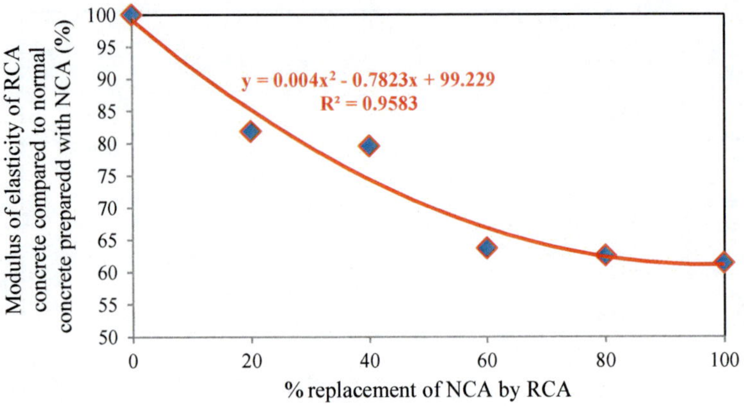

Fig. 6.8 RCA effects on the concrete elasticity modulus (Bartolacci et al. 2018)

6.6.9.1 Elasticity Modulus

The concrete elasticity modulus is increased by an aggregate of greater elastic modulus. As indicated in Fig. 6.8, the elasticity modulus of concrete thus reduces with RCA content increase in concretes. Typically, the concrete elasticity modulus having RCAs is 10–33% lesser than those of concretes of NCAs. Bartolacci et al. (2018) illustrated that the applications of 30.0% RCAs in concretes generated about 15% elastic modulus reductions. As compared to NCA concretes, the RCA concrete elastic modulus nevertheless can be lesser up to 46%. Concrete elasticity modulus reductions are because of the facts that RCAs generally have lesser elastic modulus as compared to NCAs. Also, the concrete elastic modulus reductions are related to the improved overall contents of mortars (recycled and new) that have lesser elastic moduli than most natural aggregate concretes.

6.6.9.2 Creeps and Thermal Expansions

On the RCA concrete creep, restricted studies have been conducted on. Typically, the RCA concrete creeps are greater than those of NCA concretes (Poon and Chan 2007; Anderson et al. 2009; Tabsh and Abdelfatah 2009; Andal et al. 2016). This is because creeps are relative to the contents of pastes that may be up to 51% greater in concretes of RCAs.

The thermal expansion coefficients are majorly functions of the contents and types of aggregates. Ma et al. (2019) noted that the RCA concrete thermal expansion coefficients are normally 10–30% greater than those of NCA concretes. Nevertheless, more RCA effect investigations on concrete thermal expansion behaviors are required to assert these results.

6.7 Mix Designation and Criteria-Related UPC Made with TRAC

An experimental setup is selected comprising pH meter for pH measurement, density meter for density, pin-on-disk apparatus for tribology, and modular dynamic rheometer (MCR) for rheology so as to study the drilling fluid tribology and rheology. All of the tools have their own specifications outlined by various bodies of standardization. Precautions are taken to comply with those specifications during the experiments so that findings and outcomes can be reported and compared. Professional experts and user manuals also play a significant role in carrying out experiments. While conducting measurements, both of the sources are taken into considerations.

In the fact view that acidifications are common measures for increasing productions, we conducted the assessments in 2 approaches founded on the considerations whether the weight agents are acids-soluble or acids-insoluble. The rate of damage and the flow-back pressures of both artificially fractured matrix cores and the steel cores are experimented after they have been destroyed by the TRAC. The procedures of the experiments without acid dissolutions are: displacements of kerosene, opposite displacements of oil-based fluids, displacements of paraffin. The procedures of the experiments with acid dissolutions are: displacement of kerosene-inverse oils-based boring oil displacements, in acid solutions displacements, displacements of kerosene.

Pin-on disc Tribometers from CSM Instruments are utilized in Tribology Labs so as to experiment the nano drilling fluids tribological behaviors. Along with the computer programs, the equipment enables measuring friction factors between the steel plate and the ball in the drilling fluid presence at various temperature conditions. The test starts with elimination of the formerly utilized drilling fluids from the one-side open cylinders that are furthered by drying and cleaning the inner surfaces of cylinders, the pin, and the disk. Steel ball conditions within the steel disk and pin are inspected, and essential replacements or alteration is carried out so that experiments remain dependable such as unused surface deployment of diameter of pin rotation on disks and unused ball surface for each measurement. Heating elements are also put in the cylinders to vary temperatures during the experiments, and cables are linked to the instruments so that temperatures get recorded during the experiments. Before filling the TRAC, it is often ensuring that the arm of the pin holder remains ideally horizontal during the experiments.

6.8 Specifications and Testing Standards for UPC Made with TRAC

TRAC rheologies are researched with the assistance of viscometers. Viscometers allow us to establish the drilling fluid rheological properties. It mostly comprises cylindrical bob and cup. The rotations provided to the bobs are transferred to fluids filled in the cups as the bobs are fully dipped in the fluids. These rotations generate drag forces in the drilling fluids. The fluid state prior to motions can impact on the drag forces. Viscometer is therefore utilized in determining the TRAC behaviors. Various rotational viscometer types are available that allow sampling and shearing drilling fluids for long time intervals so that equilibrium state may be attained or transient effects may be monitored. Revolving measurements may be carried out in two ways: shear rate monitored and shear stress managed. Rotations are produced by applications of constant torques to the equipment, and speed and velocity of rotations are established in the stress-controlled method that is further applied in calculating the rate of shear. Torque is established by maintaining the speed of rotation in sample in rate-controlled technique.

Instrument Lab in the Department of Petroleum Engineering also has the equipment to measure the nanodrilling fluid density meters. The procedures begin by establishing the software. The temperature is sustained at twenty degrees Celsius. The instruments have U-tube for measurements that have to be washed by acetone before beginning the experiments. The tiny nanodrilling volumes are transferred into the instruments, and density values are read on the screens after a short time.

The drilling fluid alkalinity is a significant characteristic which can affect the drilling fluid performance. The pH measurements are integrated in the study by use of pH meter of Mettler Toledo available in Instrument Lab. It is often made sure that the instruments are calibrated in current times prior to using the pH meter. The pH values for various nanodrilling fluids are obtained by dipping the electrode sensor in it for a few minutes.

The pressure cells are put in the rheometers after assembling it, and all essential wirings are done. The connections include silicone hose, pressure tubes, and temperature sensors (Pt100 Cable). So that, fluid dehydration can be reduced, and all the measurements are carried out at slightly higher pressures. In order for the rotations can be transmitted to the measuring cylinders, the magnetic couplings are then connected to the rheometers and by use computer programs, they are lowered. Appropriate DGMS programs are chosen in the software, and measurement schemes are developed in them. The schemes include measuring viscosity and shear stress at 4 various temperatures of 30 °C, 60 °C, 100 °C, and 140 °C, respectively, under monitored ranges of shear rates. The schemes give adequate time during experiments that ensure that the measurements are done after temperature of fluid has been attained to its anticipated value.

Modular Compact Rheometers (MCRs) in place the apparatus labs at Department of Petroleum Engineering, allows for researching of drilling fluid rheological behaviors various temperature and pressure conditions employing various systems of measurement. The tool is also equipped with appropriate computer software and programs to begin, monitor, and control the experiments. Double gap measuring system (DGMS) is being applied that uses component of pressure cell during measurements for measuring rheologies at different temperatures. These systems minimize the vaporization chances during experiments because of high temperatures. There, additive concentration in the fluids remains constant.

So as to start the tests, each of the fluids interacting with the pressure cell components is washed with water and desiccated to enable the contamination risks from formerly applied fluids. The pressure cells are assembled by putting the DGMS inner part inside pressure cups and with the help of special wrench and nut, tightening it. Special fluid volumes (5.8 ml as per guide manuals) are poured in the cups by use of adapters and cylinders are then put into the cups after placing them on the pressure heads. With the help of special wrench, the connections at the upper ends of the pressure heads are tightened.

Measurement schemes are developed in the software of the computer for experimenting tribological behaviors where parameter values are selected. The parameters are linear speed, distance travelled by pin, load on top of pin, and sample temperatures. Under the 10 N with linear speeds of 2.92 cm/s, friction factors for all TRAC and NCA samples are calculated at various temperatures (60 °C, 50 °C, 40 °C, and 30 °C). Room temperature and humidity are also entered before beginning any experiments. The fluids inside the sample holder of tribometer are then heated up to the model temperatures that can take a lot of time. The model holders together with the disk and fluid are continually rotating so that the rotations keep the content of fluid well spread and temperature of the fluid remain uniform. The experiments start by lowering down the fix arms into fluids, and loads are placed on the pin top after attaining the appropriate temperature. It is often ensured that the factors resulting in inaccurate results or poor measurements such as stalled equipment and noise are monitored carefully during the experiments. The experiments are repeated in case the friction factor measurements get broken up by any factor. Whether nano or base, the procedures are repeated for each fluid and curves of friction curve are obtained at various temperatures.

6.9 Improvement Methods for TRAC

6.9.1 RCA Quality Improvements

Thongkamsuk et al. (2017) tested heat treatment methods to enhance the RCA quality. The scholar noted that the RCAs were rationally comparable to the conservatively used river-dredged aggregate after thermal treatments at 800 °C. Verian et al. (2018) established that the concrete strengths were influenced by the unwashed RCAs utilized in the mixes of concretes. The strength reductions were however negated by the washed RCA use.

6.9.2 Adjustments of Water-to-Cement Ratios

Abdel-Hay (2017) reported that the correct water adjustments-to-cement ratios for the mixes of concretes might enhance the RCA concrete strengths. Abdel-Shafy and Mansour (2018) suggested higher RCA concrete cement contents and lesser water-to-cement ratios that those of RCA parent concretes to accomplish similar compressive strengths. Almeida and Cunha (2017) similarly noted that reducing the water-to-cement ratios to specific levels was greatly beneficial for RCA concretes to establish the thawing and freezing resistances equivalent to those NCA concretes.

6.9.3 Pozzolanic Material Integration

The RCA concrete durability and strength can be enhanced by utilized appropriate pozzolanic substances (Andal et al. 2016; Aliabdo et al. 2018; Arredondo-Rea et al. 2019). Arredondo-Rea et al. (2019) indicated that the applications of 65% fine granulated blast furnace slags and 30% pulverized fly ashes improved the RCA concrete compressive strengths to the control concrete cast levels with virgin granite gravels. Also, the fine granulated blast furnace slags and pulverized fuel ashes were successful in increasing the chloride ion penetration resistances into RCA concretes. It was in addition revealed that silica fumes are successful to greatly improve RCA concrete chloride penetration resistances.

6.9.4 Uses of New Mixing Techniques

Aliabdo et al. (2018) used 2-phase mixing approaches to acquire best-quality RCA concretes. The scholars utilized reclaimed aggregates treated with pozzolanic powders to enhance the RCA concrete properties. Also, Andal et al. (2016) designed 2-phase mixing techniques for fostering high-grade applications of RCA concretes. The investigators revealed that compared to NCA concretes, the 100% replacements of virgin concrete aggregates are possible by their mixing approaches to generate RCA concretes with suitable features, although the optimum scenario takes place with 20% NCA substitution. The results of strengths and slumps recommended that the new mixing methods contributed greatly to accomplish high flexural and compressive strengths and best workability. Besides, RCA concrete interfacial transition zones were realized by scanning electron microscopes (SEMs). The SEM findings established that the new techniques of mixings contributed to dense microstructures. The inner bleedings in additions can be minimized by the methods of mixings.

6.9.5 Prolonged Curing Techniques

Using a prolonged curing in wet environments is another approach of enhancing the RCA concrete performances (Bui et al. 2018; Dimitriou et al. 2018; Hou et al. 2019; Mohammed and Najim 2020). One of the most applied approaches to reduce the RCA concrete carbonation rate is extended curing. In RCA concretes, the carbonation depths are virtually double lesser when the concretes are cured with water.

6.9.6 RCA Soaking in Pozzolanic Liquids or Mix Water

The RCA concrete strengths may be boosted by either allowing reclaimed aggregates to soak up parts of mixing water without or with pozzolanic liquids during mixings or soaking the RCAs in mixes of pozzolanic liquids such as colloidal silicas or water before mixing of concretes. The micro-cracks in RCAs are expected to be filled up by the absorbed pozzolanic liquids or absorbed water with cement gels during pozzolanic reactions or cement hydrations. Therefore, the RCA concrete strengths can be increased.

6.10 Practical, Economic, and Environmental Issues of UPC Made with TRAC

Apparently, for economic feasibility in manufacturing RCA, ready-mixed concrete production factories can be utilized to get manufacturing scale to produce. An economic impact assessment is founded on an additional foundation where case action is regarded to consume no additional fee. It considers reusing concrete wastage in ready-mixed concrete production plants as infrastructural and construction resources. The developed prototype originated from information acquired by different industrial as well as concrete reprocessing workshops that are utilized as scenario research to estimate the item additional vending price. The best types of RMC manufacturing workshops to be used are the OB and FEL plants because of their aggregate feeding mechanisms. The current challenges faced in Thailand, where there are very many concrete waste disposals from destruction plans plus unmaintainable concrete being utilized, will be managed using RCA production (Rattanachu et al. 2019). The following are the summaries of conclusions and recommendations.

- The increasing replacement of RCA is primarily due to the upsurge in extra unit cost of resources. Besides, the addition of additional cement in the original case (65.00 MPa compressive power and 655.00 kg/m^3 of Category 1 cement) rises with modification rates of RCA, while in the next case (65 MPa compressive strength and 603.00 kg/m^3 of Category 1 cement), substitution of the contents of cement rises as a result of increasing the replacement rate of RCA.

- The approximated breakdown of the RCA cost is due to additional quality regulation and the value of preprocessing. However, the pre-use price could be reduced by changing the factory directly to a highly performing concrete.
- The construction apprenticeship scheme cost upsurges when produced in "overhead bin" factories and reduces when manufactured in FEL workshops. The rise in unit marginal selling cost in RACs manufactured by OB factories is primarily due to growth in the approximated UIPC.

Compared to the recycling approach in which building wastes are utilized in recycling facilities to generate new materials, the traditional costs are very high (Wijayasundara et al. 2016). The recycling process first phase involves the building wastes that are discarded in the recycling facilities at low discarding charges. Extra costs include energy consumption and transport costs. Stockpiling in which single labor is deployed at $18/h rate is the second phase. Sorting in which machines like pulverizer and excavator are used is the third phase. Also, extra costs of equipment maintenances, labor charges, fuel charges, fixed indirect costs, cost of working capital, operating charges, equipment expenses, and capital costs are charged (Wijayasundara et al. 2017). The crushing processes in fourth phase involve magnetic separations, primary grinding, and secondary crushing. It includes fuel charges, fixed indirect costs , operational costs, charges of working capital, and capital costs. Through magnetic separation, this process also involves the steel scrap sorting and is sold at around $100 per ton. The fifth phase involves the processes of manual removals in which pieces of papers, woods, and plastics are sorted out from the grounded concretes. In this process, the labor involved is charged at rates of around $18.00 per hour. The sixth phase involves air sitting, screenings, and washing that is identical to the traditional approach. Recycled fuels and waters are utilized in setting the dust particles in this stage (Wijayasundara et al. 2016). The recycling process final stages involve finished commodities that are sold at charges varying from $14.00 to $22.00.

Compared to the traditional approach, the RCAs, therefore, have the most cost-effective gains (Wijayasundara et al. 2018). While the traditional method received an adverse net benefit approximated at 44.076 billion, the recycling processes receive yearly net benefits of around $30.916 billion. The financial benefits of recycling concretes result in more long-run benefits that utilizing natural concretes.

Sensitivity analyses are carried out by putting into consideration the major parameter because of the anticipated uncertainties with the major variables utilized in the model (O'Donnell et al. 2019). The goal of the analyses is to determine the major variable effect to a modification in one unit of the output. The major factors put into consideration are the distance of the transportations, the extra RCA preprocessing costs, the extra cements needed to produce RCAs, and the compositions of RCAs (Ma et al. 2019). With depleted NAs, smashed demolition concretes are produced in large quantities and they need to be used in an ecological manner and cost-efficiently. The deterministic values are determined as the average because the parameters have normal distribution characteristics. The range of input variables can be assessed to determine the upper and lower bounds. The means and standard deviations are indicated in table nine, in which the average is a 6th of the margins. The marginal price

RCA range is also shown in table ten (Meneses et al. 2018). The marginal price RCA range is found by driving the changes in the costs of commodity by changes in the quantities.

6.11 Conclusions

There are notable potentials of reclaimed demolished concretes for applications in value-added uses in maximizing ecological and economic benefits. Through changing RCAs into meaningful resources in the new concrete generation, remarkable savings can be accomplished. The current research has critically reviewed recycled concrete aggregate to produce ultra-performance concrete structures. From this paper, the following conclusions are made:

- As NCA substitutes, RCAs are useful to generate concretes with standard durability and properties. The major challenge of applying RCAs in new concretes however their inconsistent qualities, especially when they are derived through demolitions of ancient concrete buildings.
- By meeting the virgin aggregate standard specifications, RCAs can be effectively utilized in new concretes. New guidelines and specifications however are required for RCAs.
- The RCA physical characteristics greatly affect the hardened and fresh concrete properties. The crushing values, aggregate abrasions, and impacts of RCAs influence the concrete strength features. Besides, the RCA negative chemical characteristics can influence the RCA concrete durability and therefore their performances in conditions of service.
- The 100% RCA is likely to generate concretes with standard qualities. The concretes generated with RCAs have fundamentally 81–2% of a similar NCA concrete strength. The insufficiently dense transition zones between bulk cement pastes and RCA and the adverse RCA properties cause the RCA concrete strength reductions.
- The grounded RCA applications in mixes of new concretes need cautious investigations, as the reclaimed fines further minimize the concrete strengths. Natural sand substitutions generally by recycled concrete aggregates up to about 20.1% are standard.
- The RCA concrete performances can be improved by using extended curing, utilizing new mixing methods, integrating pozzolanic substances, and modifying water-to-cement ratios.
- In best-quality concretes like self-consolidating, high-strength, and best-performance concretes, RCAs can be used by suitable mix designs and selection of materials.

The concretes having RCAs can be correctly developed and balanced to reduce the RCA impacts on its hardened and fresh durability and features, irrespective of the variations in characteristics between natural concrete aggregates and recycled concrete aggregates. The RCA chemical impurities should be reduced to promote its applications in concrete sector.

The following are the summaries of conclusions and recommendations.

- The notion about utilizing RMC production workshops to manufacture recycled concrete aggregate is misplaced since construction manager as constructor might initiate extreme additional prices as a result of particular conditions of workshops. Also, introducing FRCA during the employment of RMC production factories in RCA manufacturing is very difficult. Nanotechnology can be used to advance RCA manufacturing, because of its enormous potential of improving production and quality in highly performing construction apprenticeship scheme manufacturing. Furthermore, the use of nanotechnology focuses on the alteration of régime d'apprentissage en construction using nanomaterials. The expertise portrays several RCA properties at a nanoscale capacity utilizing different advanced equipments. Therefore, it is worth concentrating on nanoscience studies, which is crucial in the RCA microstructure research. Also, RCA's powered features can be enhanced by the use nanomaterials to obtain greater strength than National Academy of Construction.
- The inclusion of highly performing pozzolans to the reprocessed aggregate concrete to enhance RCA's performance is likely to increase the cost of utilizing it. Nonetheless, cost-benefit evaluation about the increment of fly ash indicates that it is essential because of the increase in strength. Concrete with low quality is among the burdens linked with the usage of reprocessed resources.
- To boost the implementation of recycled concrete aggregate manufacturing, its use must be introduced in government projects. Integration of advanced manufacturing technology that improves quality is essential to reduce the cost of using recycled concrete aggregate. These technologies can be used through client finance and the government, including incentives that may support the procedure. To encourage this idea, all workers should go through home training programs to facilitate the process, which is likely to improve environmental mindfulness.

References

Abdel-Hay AS (2017) Properties of recycled concrete aggregate under different curing conditions. HBRC J 13(3):271–276

Abdel-Shafy HI, Mansour MS (2018) Solid waste issue: sources, composition, disposal, recycling, and valorization. Egypt J Petrol 27(4):1275–1290

Abdulmatin A, Tangchirapat W, Jaturapitakkul C (2019) Environmentally friendly interlocking concrete paving block containing new cementing material and recycled concrete aggregate. Euro J Environ Civ Eng 23(12):1467–1484

Ahrari A, Haghani A (2019) A new decision support system for optimal integrated project scheduling and resource planning. Int J Inf Technol Project Manage 10(3):18–33

Al-Bayati HKA, Das PK, Tighe SL, Baaj H (2016) Evaluation of various treatment methods for enhancing the physical and morphological properties of coarse recycled concrete aggregate. Constr Build Mater 112:284–298

Al-Bayati HKA, Tighe SL, Achebe J (2018) Influence of recycled concrete aggregate on volumetric properties of hot mix asphalt. Resour Conserv Recycl 130:200–214

Aliabdo AA, Elmoaty AM, Fawzy AM (2018) Experimental investigation on permeability indices and strength of modified pervious concrete with recycled concrete aggregate. Constr Build Mater 193:105–127

Almeida A, Cunha J (2017) The implementation of an activity-based costing (ABC) system in a manufacturing company. Proc Manuf 13:932–939

Al-Swaidani A M, Khwies WT (2018) Applicability of artificial neural networks to predict the mechanical and permeability properties of volcanic scoria-based concrete. Adv Civ Eng

Andal J, Shehata M, Zacarias P (2016) Properties of concrete containing a recycled concrete aggregate of preserved quality. Constr Build Mater 125:842–855

Anderson KW, Uhlmeyer JS, Russell M (2009) Use of recycled concrete aggregate in PCCP: literature search. In: WSDOT research report WA-RD 726.1. Washington State Department of Transportation (WSDOT), Olympia, WA, 37

Arredondo-Rea SP, Corral-Higuera R, Gómez-Soberón JM, Gámez-García DC, Bernal-Camacho JM, Rosas-Casarez CA, Ungsson-Nieblas MJ (2019) Durability parameters of reinforced recycled aggregate concrete: case study. Appl Sci 9(4):617

Aslani F, Ma G, Wan DLY, Muselin G (2018) Development of high-performance self-compacting concrete using waste recycled concrete aggregates and rubber granules. J Clean Prod 182:553–566

Babel S, Ta AT, Liyanage TUH (2020) Current situation and challenges of waste management in Thailand. In: Sustainable waste management challenges in developing countries. IGI Glob:409–440

Balan MA, Thomas CA (2019) Experimental study and optimization of best performance self-compacting recycled aggregate concrete. In: National conference on structural engineering and construction management. Springer, Cham

Bartolacci F, Paolini A, Quaranta AG, Soverchia M (2018) Assessing factors that influence waste management financial sustainability. Waste Manage 79:571–579

Bui NK, Satomi T, Takahashi H (2018) Mechanical properties of concrete containing 100% treated coarse recycled concrete aggregate. Constr Build Mater 163:496–507

Chatterjee A, Sui T (2019) Alternative fuels–effects on clinker process and properties. Cem Concr Res 123:105777

Deakins D, Bensemann J (2019) Achieving innovation in a lean environment: how innovative small firms overcome resource constraints. Int J Innov Manage 23(4):1950037

Di Maria A, Eyckmans J, Van Acker K (2018) Downcycling versus recycling of construction and demolition waste: combining LCA and LCC to support sustainable policy making. Waste Manage 75:3–21

Dimitriou G, Savva P, Petrou MF (2018) Enhancing mechanical and durability properties of recycled aggregate concrete. Constr Build Mater 158:228–235

Ding Z, Yi G, Tam VW, Huang T (2016) A system dynamics-based environmental performance simulation of construction waste reduction management in China. Waste Manage 51:130–141

Du Z, Lin B (2018) Analysis of carbon emissions reduction of China's metallurgical industry. J Clean Prod 176:1177–1184

Durdyev S, Omarov M, Ismail S (2017) Causes of delay in residential construction projects in Cambodia. Cog Eng 4(1):1291117

Forsyth T (2018) Encyclopedia of international development. Routledge

Fraile-Garcia E, Ferreiro-Cabello J, López-Ochoa LM, López-González LM (2017) Study of the technical feasibility of increasing the amount of recycled concrete waste used in ready-mix concrete production. Materials 10(7):817

Guo H, Shi C, Guan X, Zhu J, Ding Y, Ling TC, Wang Y (2018) Durability of recycled aggregate concrete–a review. Cem Concr Compos 89:251–259

Hou Y, Ji X, Su X (2019) Mechanical properties and strength criteria of cement-stabilised recycled concrete aggregate. Int J Pavement Eng 20(3):339–348

Jin R, Li B, Zhou T, Wanatowski D, Piroozfar P (2017) An empirical study of perceptions towards construction and demolition waste recycling and reuse in China. Resour Conserv Recycl 126:86–98

Júnior NSA, Silva GA, Dias CM, Ribeiro DV (2019) Concrete containing recycled aggregates: Estimated lifetime using chloride migration test. Constr Build Mater 222:108–118

Ma J, Sun D, Pang Q, Sun G, Hu M, Lu T (2019) Potential of recycled concrete aggregate pretreated with waste cooking oil residue for hot mix asphalt. J Clean Prod 221:469–479

Meneses EJ, Gaussens M, Jakobsen C, Mikkelsen PS, Grum M, Vezzaro L (2018) Coordinating rule-based and system-wide model predictive control strategies to reduce storage expansion of combined urban drainage systems: the case study of Lundtofte, Denmark. Water 10(1):76

Mohammed SI, Najim KB (2020) Mechanical strength, flexural behavior and fracture energy of recycled concrete aggregate self-compacting concrete. Structures 23:34–43

O'Donnell BT, Ives CJ, Mohiuddin OA, Bunnell BA (2019) Beyond the present constraints that prevent a wide spread of tissue engineering and regenerative medicine approaches. Front Bioeng Biotechnol 7:95

Poon CS, Azhar S, Kou SC (2003) Recycled aggregates for concrete applications. In: Proceedings of the materials science and technology in engineering conference-now. New and Next, Hong Kong, China, 2003

Poon CS, Chan D (2007) Effects of contaminants on the properties of concrete paving blocks prepared with recycled concrete aggregates. Constr Build Mater 21(1):164–175

Poon CS, Kou SC, Lam L (2002) Use of recycled aggregates in molded concrete bricks and blocks. Constr Build Mater 16(5):281–289

Poon CS, Shui ZH, Lam L, Fok H, Kou SC (2004) Influence of moisture states of natural and recycled aggregates on the slump and compressive strength of concrete. Cem Concr Res 34(1):31–36

Ratnayake RC, Samarakoon SS (2017) Structural integrity assessment and control of ageing onshore and offshore structures. Model Simul Techn Struct Eng 2017:445–476

Rattanachu P, Tangchirapat W, Jaturapitakkul C (2019) Water permeability and sulfate resistance of eco-friendly high-strength concrete composed of ground bagasse ash and recycled concrete aggregate. J Mater Civ Eng 31(6):04019093

Tabsh SW, Abdelfatah AS (2009) Influence of recycled concrete aggregates on strength properties of concrete. Constr Build Mater 3(2):1163–1167

Teh SH, Wiedmann T, Moore S (2018) Mixed-unit hybrid life cycle assessment applied to the recycling of construction materials. J Econ Struct 7(1):13

Thongkamsuk P, Sudasna K, Tondee T (2017) Waste generated in high-rise buildings construction: a current situation in Thailand. Energy Proc 138:411–416

Verian KP, Ashraf W, Cao Y (2018) Properties of recycled concrete aggregate and their influence in new concrete production. Resour Conserv Recycl 133:30–49

Wang H, Sun X, Wang J, Monteiro PJ (2016) Permeability of concrete with recycled concrete aggregate and pozzolanic materials under stress. Materials 9(4):252

Wijayasundara M, Mendis P, Crawford RH (2017) Methodology for the integrated assessment on the use of recycled concrete aggregate replacing natural aggregate in structural concrete. J Clean Prod 166:321–334

Wijayasundara M, Mendis P, Crawford RH (2018) Integrated assessment of the use of recycled concrete aggregate replacing natural aggregate in structural concrete. J Clean Prod 174:591–604

Wijayasundara M, Mendis P, Zhang L, Sofi M (2016) Financial assessment of manufacturing recycled aggregate concrete in ready-mix concrete plants. Resour Conser Recycl 109:187–201

Index

N. Makul, *Principles of Cement and Concrete Composites*, Structural Integrity 18,
https://doi.org/10.1007/978-3-030-69602-3

B

J

K

L

S